Engineering Cyber-Physical Systems and Critical Infrastructures

Volume 2

Series Editor

Fatos Xhafa, Departament de Ciències de la Computació, Technical University of Catalonia, Barcelona, Spain

The aim of this book series is to present state of the art studies, research and best engineering practices, real-world applications and real-world case studies for the risks, security, and reliability of critical infrastructure systems and Cyber-Physical Systems. Volumes of this book series will cover modelling, analysis, frameworks, digital twin simulations of risks, failures and vulnerabilities of cyber critical infrastructures as well as will provide ICT approaches to ensure protection and avoid disruption of vital fields such as economy, utility supplies networks, telecommunications, transports, etc. in the everyday life of citizens. The intertwine of cyber and real nature of critical infrastructures will be analyzed and challenges of risks, security, and reliability of critical infrastructure systems will be revealed. Computational intelligence provided by sensing and processing through the whole spectrum of Cloud-to-thing continuum technologies will be the basis for real-time detection of risks, threats, anomalies, etc. in cyber critical infrastructures and will prompt for human and automated protection actions. Finally, studies and recommendations to policy makers, managers, local and governmental administrations and global international organizations will be sought.

Issa Traore · Isaac Woungang · Sherif Saad
Editors

Artificial Intelligence for Cyber-Physical Systems Hardening

Springer

Editors
Issa Traore
Department of Electrical and Computer Engineering
University of Victoria
Victoria, BC, Canada

Isaac Woungang
Department of Computer Science
Ryerson University
Toronto, ON, Canada

Sherif Saad
School of Computer Science
University of Windsor
Windsor, ON, Canada

ISSN 2731-5002 ISSN 2731-5010 (electronic)
Engineering Cyber-Physical Systems and Critical Infrastructures
ISBN 978-3-031-16236-7 ISBN 978-3-031-16237-4 (eBook)
https://doi.org/10.1007/978-3-031-16237-4

This Springer imprint is published by the registered company Springer Nature Switzerland AG
The registered company address is: Gewerbestrasse 11, 6330 Cham, Switzerland

To Our Families

Preface

The merging of computer systems and physical–mechanical systems led to the creation of cyber-physical systems (CPSs). A cyber-physical system combines software systems, sensors, and actuators connected over computer networks. Through these sensors, data about the physical world can be captured for processing. The software system processes the captured data from the sensors and makes decisions. The decisions are sent as signals to trigger the actuators to perform specific actions that affect the state of the physical world. Today examples of cyber-physical systems are everywhere, from autonomous vehicles to wearable devices.

The fact that CPS directly impacts the physical world made the security, privacy, and trust of cyber-physical systems a significant concern. The complexity and the scale of CPS introduced various security, privacy, and trust challenges that traditional security engineering practices cannot handle.

The application of artificial intelligence (AI) for cyber-physical systems hardening encompasses various techniques, methodologies, and best practices to mitigate vulnerabilities and security risks in CPS by eliminating potential security threats and reducing the systems' attack surface. The book focuses on the applications of artificial intelligence in hardening CPS against conventional and unconventional attack vectors. It covers diverse methods and techniques for hardening software, hardware, firmware, infrastructure, and communication channels in CPS. Many researchers and professionals from academia and industry have contributed theoretical and applied studies and real-world case studies to this book.

This book consists of ten chapters organized as follows.

The first chapter is an "Introduction" to cyber-physical systems by the editors. It defines cyber-physical systems, gives an overview of their underlying characteristics and design goals, and discusses barriers to developing intelligent hardening systems.

In the second chapter entitled "Machine Learning Construction: Implications to Cybersecurity", by Waleed A. Yousef, the author focuses on the importance of machine learning (ML) in the field of cyber-physical security, in the sense that ML can enable the design of detection algorithms that have the capability of learning from security data to hunt several kinds of threats, achieve better monitoring, master the complexity of the threat intelligence feeds, and achieve timely remediation of security incidents. The field of ML can be decomposed into two basic subfields, namely construction and assessment. In the one hand, construction refers to designing or inventing an appropriate algorithm that learns from the input data and achieves a good performance according to some optimality criterion. On the other hand, the assessment part consists in attributing some performance measures to the constructed ML algorithm, along with their estimators, to objectively evaluate the algorithm. This chapter focuses on the basics of the construction part, emphasizing the associated mathematical foundations and underlying concepts. The author first reviews some of the regression and classification methods used for predicting a quantitative or categorical response variable, respectively. Then, the basic concepts related to the performance of these methods are presented.

In the third chapter entitled "Machine Learning Assessment: Implications to Cybersecurity", by Waleed A. Yousef, the author focuses on assessment and performance estimation of ML algorithms, in the context of cybersecurity and cyber-physical security design. In this chapter, a comprehensive review of nonparametric methods to estimate a statistic from just one available dataset through resampling techniques, along with the literature to establish a coherent theoretical framework for these methods, which can be used to estimate the error rate (a one-sample statistic) and the area under the ROC curve (AUC) (a two-sample statistic), is presented.

In the fourth chapter entitled "A Collection of Datasets for Intrusion Detection in MIL-STD-1553 Platforms", by Hadeer Saad, Issa Traore, Paulo Quinan, Karim Ganame, and Oussama Boudar, the authors focus on the security concerns of the MIL-STD-1553, a military standard communication protocol for the operation of a wide range of defense platforms. From a cybersecurity perspective, the inherent vulnerabilities in MIL-STD-1553 data buses represent prime conduits for compromising defense platforms that rely on them for communications. This chapter explores a range of cyber-attacks against MIL-STD-1553 data buses and presents a collection of datasets that were generated by executing selected attack scenarios in a testbed environment. These datasets can be used toward designing and evaluating intrusion detection systems for MIL-STD-153 avionic platforms. The authors first review the key concepts and requirements that underlie the normal operation of the mil-std-1553 standard, which has led to a foundation to define a baseline model for the normal operation and behavior of MIL-STD-1553 systems. Next, different potential attack vectors against MIL-STD-1553 are discussed, along with the different attack scenarios that can be used for evaluating an MIL-STD-1553 IDS. Finally, the simulation environment, procedures, and scenarios used to generate the datasets, are described. This dataset is available at the Information Security and Object Technology (ISOT) Lab, University of Victoria, BC, https://www.uvic.ca/ecs/ece/isot/datasets/index.php.

In the fifth chapter entitled "Unsupervised Anomaly Detection for MIL-1553 Avionic Platform Using Cusum", by Krunal Sachdev, Hadeer S. Ahmed, Issa Traore, Karim Ganame, and Oussama Boudar, the authors continue the investigation started in the fourth chapter of MIL-STD-1553, a military standard developed by the US department of defense for communication among the military avionic platforms.

They propose an unsupervised anomaly detection scheme using the CUSUM algorithm for the MIL-STD-1553 protocol. Then, from the dataset constructed in the fourth chapter, a set of relevant ML features were extracted based on leveraging the time-based properties of the communication bus. These features were then fed to the CUSUM algorithm for detection purpose. The experimental evaluation of the proposed detector using that dataset showed promising results.

In the sixth chapter entitled "Secure Design of Cyber-Physical Systems at the Radio Frequency Level: Machine and Deep Learning-Driven Approaches, Challenges, and Opportunities", by Ceren Comert, Omer Melih Gul, Michel Kulhandjian, Azzedine Touazi, Cliff Ellement, Burak Kantarci, and Claude D'Amours, the authors focus on the issue of deploying new 5G services on many of the critical infrastructures such as connected vehicles, remote health care and smart infrastructures on radio frequency (RF)-based networks, and the ability to protect these new wireless networks and the radio spectrum. They put forward the use of artificial intelligence (AI)-based transmitter fingerprinting as an efficient solution to identify and track the unintended interference sources or malicious actors. As new automobiles are expected to be equipped with vehicle to infrastructure (V2I), vehicle to vehicle (V2V), and other telecommunications capabilities, the AI-based technique and other new automated ones are needed to protect connected and autonomous vehicles (CAVs) on the road from unintentional or malicious interference. This chapter presents the state of the art in real-time decision support systems for the cyber-physical systems that build on critical infrastructures such as CAVs, through radio fingerprinting solutions. The authors first present the legacy approaches used to detect, classify, and identify a transmitter. Next, machine and deep learning-based (ML/DL) approaches for transmitter identification using RF fingerprinting techniques are discussed. Finally, a comparative study on the open issues, challenges, and opportunities toward ML/DL-driven security of the critical cyber-physical systems through RF fingerprinting is presented.

In the seventh chapter entitled "Attack Detection by Using Deep Learning for Cyber-Physical System", by Saeid Jamshidi, Amin Nikanjam, Mohammad Adnan Hamdaqa, Foutse Khomh, the authors address the issue of detecting cyber-attacks on cyber-physical system (CPS), stressing on the need to have the CPS security measures implemented. This chapter discusses the various deep learning (DL) and reinforcement learning (RL) models to address the detection of cyber-attacks in CPS. Challenges to attack detection in CPS are discussed, along with common datasets used for DL in CPSs. The authors first present an overview of DL in CPSs. Next, RL and DRL techniques that have been successfully utilized in the field of cybersecurity/CPSs are presented, and then the challenges to attack detection in CPSs and some robust attacks detection schemes that are available in the literature are discussed.

In the eighth chapter entitled "Security and Privacy of IoT Devices for Aging in Place", by Noel Khaemba, Issa Traoré, and Mohammad Mamun, the authors investigate the issue of rising cost of elderly living and care facilities, and report on the need for smart home solutions involving the use of emerging technologies centered around smart IoT devices from a security standpoint. More precisely, to ensure security and privacy for a smart home for aging in place, different aspects of the IoT devices must be considered. This chapter seeks to provide a categorical review and analysis of agetech IoT device technologies. The underlying security and privacy challenges and available solutions are discussed in-depth. Indeed, starting with a discussion on the major categories of devices used for aging in place (AIP) and the corresponding use cases, the common threats and vulnerabilities faced by AIP technologies and threats common to the broad IoT user population are discussed. Finally, the relevance of using artificial intelligence to tackle agetech security is discussed and the relevant datasets and approaches are described.

In the ninth chapter entitled "Detecting Malicious Attacks Using Principal Component Analysis in Medical Cyber-Physical Systems", by Wei Lu, the author raises attention to the surge of cybercriminal activities targeting the medical cyber-physical systems from a security and privacy perspectives. Existing security solutions in this domain are mainly prevention-based and are highly insufficient due to the power consumption and costly resources when implementing computationally expensive solutions. This chapter proposes an anomaly detection system based on the principal component analysis to assure the security of networked medical devices. The proposed approach is evaluated by considering a publicly available dataset collected in a real-time medical cyber-physical system testbed network, showing promising results in terms of detection of malicious attacks with a high detection rate and an acceptable low false alarm rate. The authors first present the concept of intrusion detection systems, followed by the proposed principal component analysis (PCA)-based anomaly detection scheme, and ending with its performance evaluation study.

In the tenth chapter entitled "Activity and Event Network Graph and Application to Cyber-Physical Security", by Paulo Gustavo Quinan, Issa Traore, and Isaac Woungang, the authors introduce the activity and event network (AEN) as a new large graph model that enables describing and analyzing continuously in real-time the key security-relevant information about the operations of networked systems and data centers. This model allows identifying long-term and stealthy attack patterns, which may be difficult to capture using traditional approaches. The focus of the chapter is on defining the model elements and the underlying graph construction algorithms. The author first defines the theoretical foundation of the AEN graph model, then presents the data sources used to construct the graph. Next, the AEN graph model elements are defined by presenting the node and edge types that are involved. Following these concepts, the AEN underlying probability model framework architectures are presented. Finally, an illustrative case study based on an existing cyber-physical security dataset is described in-depth.

The guest editors of this book wish to thank the contributing authors for their interesting contributions, as well as for their timely collaboration in the preparation of their chapters. We also would like to appreciate the anonymous reviewers for their careful reviews of the contributed chapters, and their useful suggestions and feedback that helped the authors improve the quality of their manuscripts. Finally, we would like to sincerely thank Springer for this opportunity.

Victoria, BC, Canada — Issa Traore
Toronto, ON, Canada — Isaac Woungang
Windsor, ON, Canada — Sherif Saad

Acknowledgements We would like to thank all the authors who have contributed quality chapters to this book. Special thanks to all our editorial advisory board members and the reviewers who invested a lot of efforts and time in selecting the highest quality chapters possible. We would like also to thank the *Springer* team who helped and advised us in conducting this book project to term.

Finally, we would like to thank our families for their tireless support throughout this project.

Contents

Introduction

Issa Traore, Isaac Woungang, and Sherif Saad

Abstract Cyber-physical systems (CPS) arise from the merging of computer systems and physical–mechanical systems. Although they build on existing proven technologies (e.g., control systems, wireless sensor networks, cyber systems) as a whole, they represent an emerging technology that is poised to have disruptive and consequential impact in the next few years. This chapter provides a general overview of CPS and discuss the underlying challenges and opportunities.

Keywords Artificial intelligence · Cyber-physical systems · Vulnerabilities · Security risks · Conventional attack vectors · Unconventional attack vectors · Hardware · Firmware · Infrastructure · Communication channels

1 Context and Definition

Cyber-physical systems bridge the gap between physical and virtual components. It is a growing technology which merges computational and physical components and span many industries. Many of these industries play a central role in mission-critical and life-critical sectors such as smart grids, nuclear power plants, smart manufacturing, automated transportation and vehicles, robotic surgery, personalized healthcare, smart building, smart agriculture, and supply chain automation and control, to name a few.

While physical systems have been around for a very long time, automating tasks and processes to achieve autonomy in the systems has lagged [1]. However, recent

I. Traore (✉)
Department of Electrical and Computer Engineering, University of Victoria, Victoria, BC, Canada
e-mail: itraore@ece.uvic.ca

I. Woungang
Department of Computer Science, Toronto Metropolitan University, Toronto, ON, Canada
e-mail: iwoungan@scs.ryerson.ca

S. Saad
School of Computer Science, University of Windsor, Windsor, ON, Canada
e-mail: Sherif.Saad@uwindsor.ca

I. Traore et al. (eds.), *Artificial Intelligence for Cyber-Physical Systems Hardening*, Engineering Cyber-Physical Systems and Critical Infrastructures 2,
https://doi.org/10.1007/978-3-031-16237-4_1

advances in wireless sensor network (WSN), cloud computing, machine learning and deep learning, are enabling efficient and effective integration of sensing, processing, and networking, to deliver automation of CPS processes across many industries.

As an emerging technology, CPS brings new opportunities, but also faces many challenges [2]. Challenges include, among others, mastering the heterogeneity and complexity of the components involved, addressing integrability and interoperability requirements, addressing the need for real-time data processing and fast data communication, ensuring safety, and data and communication security.

2 Characteristics and Design Goals

Major components of typical CPS include sensors, actuators, and computation modules, interconnected through network artifacts. To achieve the goal of CPS, the integration of the components combines feedback control, fast data transfer and processing, and intelligent computation.

According to Lozano and Vijayan [1], the required characteristics of a CPS include autonomy, stability, robustness, efficiency, scalability, safety, reliability, accuracy, and connectivity. A key characteristic that is missing from this list and should be added is security.

One of the most prominent among these characteristics is autonomy, which refers to the decision-making ability of a CPS. While autonomy can be implemented in many ways, using artificial intelligence (AI) and machine learning (ML) techniques provide an edge in achieving such goal. However, the imperfections inherent in AI and ML techniques pose some challenges in achieving characteristics such as stability, reliability, and accuracy. When leveraging AI/ML techniques to achieve autonomy, trade-offs must be done to deliver the most accurate, reliable, and stable system.

3 Security and Hardening

As CPS integrates the physical world and cyberspace, Jamal et al. [3] advocates the need for an integrated approach in hardening CPS, where security concerns are addressed in both physical components (e.g., embedded controllers) and cyberspace components (e.g., digital systems) conjointly rather than separately as it has been the case traditionally for many systems. The hybrid automation architecture proposed by Tantawy et al. [4] is an example of such integrated approach. The proposed approach relies on a tree-based model that uses the same data for both process automation and attack identification.

While many traditional cyberspace attacks are still largely applicable to CPS (e.g., denial of service (DOS), man-in-the-middle (MITM), replay attacks, deception attacks), several new attacks specific to physical aspects of CPS have emerged such as time synchronization attacks, jamming attacks, sophisticated data poisoning and injection attacks (e.g., on compressive sensing), primary user (PU) emulation attacks, device cloning and SIM swapping attacks, to name a few.

Such threats originate both from external adversaries seeking to exploit vulnerabilities in cyberspace systems and physical processes as well insiders leveraging their inside knowledge and privileges to achieve the same goals. In this context, in addition to adopting an integrated design approach, hardening strategies must identify the interdependencies between cyber and physical components and the interactions between cyber-attackers and system operators [5]. This would provide the foundation for robust and sound security risk identification, mitigation, and monitoring for CPS [6].

4 Intelligence

The last two decades have witnessed a tremendous progress in AI, ML, and deep learning techniques, with the appearance of new algorithms and techniques to address the challenges that hamper accuracy and precision (e.g., noisy data, missing data, concept drift), processing speed and data storage, and data privacy, trust, and security. New techniques and architectures have appeared (e.g., federated learning, distributed ML, meta learning, distributed control systems and networked control systems) that help assuage these concerns [7].

Considering that many security systems deal with imperfect data and are themselves inherently pattern recognition systems, there is also a rich tradition of using AI and ML techniques to design such systems.

Not surprisingly, the hardening of CPS is leveraging the knowledge and background accumulated through the years in advancing AI and ML and developing intelligent cybersecurity systems which have mastered how to handle many threats common to both traditional cyber systems and emerging CPS. However, despite the overlaps in attack vectors between CPS and cyber systems, CPS have unique characteristics that should be leveraged in building intelligent hardening systems. For instance, there is still a strong tendency in many existing CPS to depend heavily or solely on traditional cyberspace datasets such as network traffic. For instance, most existing IoT security datasets cover only attack data based on network traffic and involve only traditional attack vectors. On the other hand, most CPS datasets (such as smart home datasets) cover only legitimate/normal occurrences or operations, without any actual attack traces, which make the design of intelligent hardening systems dependent on only synthetically generated attack samples. The challenges in generating intrinsic CPS operational and environmental attack samples are a tremendous barrier toward developing hybrid automation systems that take advantage of the intrinsic process automation data for intelligent system hardening.

Cyberphysical systems are inherently multidisciplinary. Building intelligent hardened CPS also involves multiple disciplines such as control theory and engineering, communication engineering and networking, cybersecurity, artificial and machine intelligence, system engineering, etc. Finding a single individual with expertise in all these disciplines is almost impossible or very rare. Therefore, multidisciplinary teams with expertise in the different areas involved are likelier to achieve better success in the design and development of such systems. Specifically, to avoid mistakes made previously in the design of intelligent security systems such as ML/AI-based intrusion detection systems (IDSs) [8, 9], it is important to ensure that the right expertise is available when designing intelligent hardened CPS. For instance, this requires strong expertise in security threat and vulnerabilities identification modelling, and a deep understanding of machine learning *construction* and *assessment* fundamentals and techniques and its implication for system security.

5 Summary

Like any emerging technology, CPS are full of opportunities, but also faces great challenges. These challenges must be addressed for CPS to fully achieve its promises. Some of these challenges arise from the security, privacy and trust issues created by the interconnection between the physical components and structures which used to be deployed and operate in isolation in closed loop, with the cyberspace, which inherently is open.

The remaining chapters in this book discuss some of the challenges involved in hardening CPS using intelligent models and present some solutions toward achieving such goal. Specifically,

- Chapters 2 and 3 explore machine learning model construction and assessment and the implications for cybersecurity, which is a core aspect of cyberphysical system security.
- Data is the bloodline of machine and deep learning. Chapter 4 tackles the need for adequate datasets in building hardened CPS, by presenting a collection of IDS datasets for the MIL-STD-1153 protocol, which is a core protocol underlying data buses interconnecting avionics systems in military aircraft.
- Chapters 6 and 7 discuss the use of machine and deep learning for secure design of CPS at the radio frequency level and for attack detection in these systems.
- CPS involve different applications such as smart healthcare, smart manufacturing, and smart cities [2]. Chapters 5, 8, and 9 present concrete intelligent hardening models for CPS in different application areas, including ageing in place in smart homes, avionics data buses, and networked medical devices.
- Chapter 10 presents a new graph-based framework to model and detect long-term threats, along with a case study based on an existing IoT security dataset.

Although these chapters cover only a few aspects of intelligent CPS hardening, they help advance and improve our understanding of the challenges and opportunities involved in developing and maintaining secure CPS architectures.

References

1. Lozano CV, Vijayan KK (2020) Literature review on cyber physical systems design. In: 10th Conference on learning factories, CLF2020, Procedia manufacturing, vol 45. Elsevier, pp 295–30, 205–300
2. Aguida MA, Ouchani S, Benmalek M. A review on cyber-physical systems: models and architectures. In: Proceedings 2020 IEEE 29th international conference on enabling technologies: infrastructure for collaborative enterprises (WETICE), pp 275–278
3. Jamal AA, Majid A-AM, Konev A, Kosachenko T, Shelupanov A. A review on security analysis of cyber physical systems using Machine learning. Mater Today: Proc. https://doi.org/10.1016/j.matpr.2021.06.320 (Elsevier)
4. Tantawy A, Abdelwahed S, Erradi A, Shaban K (2020) Model-based risk assessment for cyber physical systems security. Comput Sec 96:101864. https://doi.org/10.1016/j.cose.2020.101864
5. Karangelos E, Wehenkel L (2022) Cyber–physical risk modeling with imperfect cyber-attackers. Electric Power Syst Res 211:108437
6. Jbair M, Ahmad B, Maple C, Harrison R (2022) Threat modelling for industrial cyber physical systems in the era of smart manufacturing. Comput Ind 137:103611 (Elsevier)
7. Tahoun AH, Arafa M (2021) Secure control design for nonlinear cyber-physical systems under DoS, replay, and deception cyber-attacks with multiple transmission channels. ISA Trans. https://doi.org/10.1016/j.isatra.2021.11.033
8. McHugh J (2000) Testing intrusion detection systems: a critique of the 1998 and 1999 DARPA intrusion detection system evaluations as performed by Lincoln Laboratory. ACM Trans Inf Syst Secur 3(4):262–294. https://doi.org/10.1145/382912.382923
9. McHugh J (2001) Intrusion and intrusion detection. Int J Inf Secur (IJIS 1:14–35. https://doi.org/10.1007/s102070100001 (Springer)

Machine Learning Construction: Implications to Cybersecurity

Waleed A. Yousef

Abstract Statistical learning is the process of estimating an unknown probabilistic input-output relationship of a system using a limited number of observations. A statistical learning machine (SLM) is the algorithm, function, model, or rule, that learns such a process; and machine learning (ML) is the conventional name of this field. ML and its applications are ubiquitous in the modern world. Systems such as Automatic target recognition (ATR) in military applications, computer aided diagnosis (CAD) in medical imaging, DNA microarrays in genomics, optical character recognition (OCR), speech recognition (SR), spam email filtering, stock market prediction, etc., are few examples and applications for ML; diverse fields but one theory. In particular, ML has gained a lot of attention in the field of cyberphysical security, especially in the last decade. It is of great importance to this field to design detection algorithms that have the capability of learning from security data to be able to hunt threats, achieve better monitoring, master the complexity of the threat intelligence feeds, and achieve timely remediation of security incidents. The field of ML can be decomposed into two basic subfields: *construction* and *assessment*. We mean by *construction* designing or inventing an appropriate algorithm that learns from the input data and achieves a good performance according to some optimality criterion. We mean by *assessment* attributing some performance measures to the constructed ML algorithm, along with their estimators, to objectively assess this algorithm. *Construction* and *assessment* of a ML algorithm require familiarity with different other fields: probability, statistics, matrix theory, optimization, algorithms, and programming, among others. To help practitioners, specially those of cyberphysical security, to understand the theoretical foundations of ML, before they delve into whole books, we compile the very basics of the first of these two subfields (*construction*) in this chapter. In addition to explaining the mathematical foundations of the field, we emphasize the intuitive explanation and concepts.

W. A. Yousef (✉)
CS Department, HCILAB, Faculty of Computers and Artificial Intelligence, Helwan University, Helwan, Egypt
e-mail: wyousef@fci.helwan.edu.eg

I. Traore et al. (eds.), *Artificial Intelligence for Cyber-Physical Systems Hardening*, Engineering Cyber-Physical Systems and Critical Infrastructures 2,
https://doi.org/10.1007/978-3-031-16237-4_2

Keywords Statistical learning · Machine learning · Construction · Sample · Learning · Prediction · Regression · Classification · Quantitative variable · Categorical variable · Performance · Statistical decision theory

1 Introduction

1.1 Motivation

Consider a sample consisting of a number of cases (observations), where each case is composed of a set of inputs and the corresponding output, all of which will be given to a learning algorithm. Such a sample provides the means for the algorithm to learn during its so-called *training* (or learning) stage. The goal of this training or learning stage is to understand as much as possible how the output is related to the inputs in these observations, so that when a new set of inputs is given, in the future, the algorithm will have some means of predicting the corresponding output. The above terminology has been borrowed from the field of ML. However, the roots of this problem exists originally in the field of statistical decision theory, where the terminology is somewhat different. In the latter field, the inputs are called the predictors and the output is called the response. When the output is quantitative the learning algorithm is called regression; when the output is categorical or ordered categorical the learning algorithm is called classification. In other communities, the terms *input features* and *output class* are used, respectively. The learning process can be defined as follows.

Definition 1 Learning is the process of estimating an unknown input-output dependency or structure of a system using a limited number of observations [10]. □

Statistical learning is crucial to many applications. For example, In cyberphysical security, a network activity must be classified as normal or malicious to avoid any potential threat [30]. This is an example of prediction, regardless of whether it is done by a network analyst or by a ML algorithm. In either case, the prediction is done based on learning from previous network traffics. The features, i.e., predictors, in this case may be the activity's IP address, number of scanned ports, duration of connection, etc. The output in this case, i.e., response, is categorical and belongs to the set: $\mathscr{G} = \{normal,\ malicious\}$. There are so many such examples, including email filtering and spam detection, fraud detection in financial transactions, etc. All of these examples involve a prediction step based on previous learning.

This chapter reviews some of the regression and classification methods used for predicting a quantitative or categorical response variable, respectively. In addition, the chapter explains basic concepts related to the performance of these methods. The purpose is not to present a survey as much as to introduce the field in an approach that combines both mathematics and intuition, and to explain how the different ingredients relate to each other. We hope this chapter helps practitioners realize the importance of being equipped with the minimum amount of theory before diving deeply into practice.

1.2 Notation

Some basic concepts and terminology, necessary for the sequel, must be formally introduced. The world of variables can be categorized into two categories: deterministic variables and random variables. A deterministic variable takes a definite value; the same value will be the outcome if the experiment that yielded this value is rerun. On contrary, a random variable is a variable that takes a non-definite value with a probability value.

Definition 2 A random variable X is a function from a sample space S into the real numbers $\Re$, that associates a real number, $x = X(s)$, with each possible outcome $s \in S$. □

Details on the topic can be found in [9, Chap. 1]. For more rigorous treatment of random variables based on measure theoretic approach see [4]. Variables can be categorized as well, based on value, into: quantitative (or metric), qualitative (or categorical), and ordered categorical. A quantitative variable takes a value on $\Re$, and it can be discrete or continuous. A categorical variable does not necessarily take a numerical value; rather it takes a value from a finite set. E.g., the set $\mathscr{G} = \{red,\ green,\ blue\}$ is a set of possible qualitative values that can be assigned to a color. An ordered categorical variable is a categorical variable with relative algebraic relations among the values. E.g., the set $\mathscr{G} = \{small,\ medium,\ large\}$ includes ordered categorical values.

Variables in a particular process are related to each other in a certain manner. When variables are random the process is said to be stochastic, i.e., when the inputs of this process have some specified values there is no deterministic value for the output, rather a probabilistic one. The output in this case is a random variable.

Before delving into mathematical details, it is convenient to introduce some commonly used notation. A random variable—or a random vector—is referred to by an upper-case letter, e.g., X. An instance, case, or observation, of that variable is referred to by a lower-case letter, e.g., x. A collection of n observations for the p-dimensional random vector X is collected into an $n \times p$ matrix and represented by a bold upper-case $\mathbf{X}$. A lower-case bold letter $\mathbf{x}$ is reserved for describing a vector of any n-observations of a variable, even a tuple consisting of non-homogeneous types. The main notation in the sequel will be as follows: $\mathbf{tr} : \{t_i = (x_i, y_i)\ ,\ i = 1, \ldots, n\}$ represents an n-case training dataset, i.e., one on which the learning mechanism will execute to train, or learn. Every observation t_i of this set represents a tuple of the predictors x_i represented in a p-dimensional vector, and the corresponding response variable y_i. All the n observations x_i's may be written in a single $n \times p$ matrix $\mathbf{X}$, while all the observations y_i may be written in a vector $\mathbf{y}$. Some terminologies

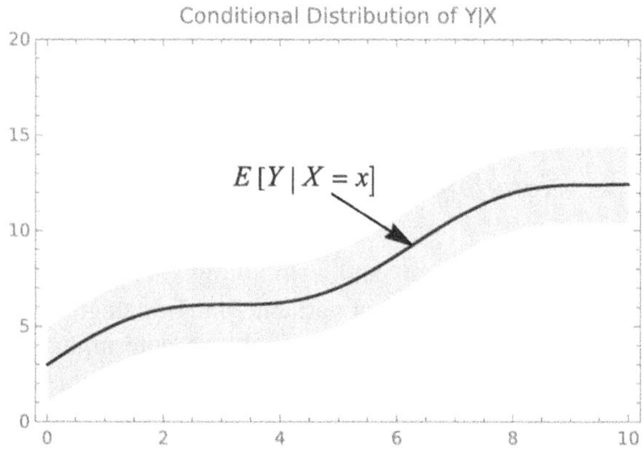

Fig. 1 Conditional expectation of a r.v. Y, conditional on a r.v. X, is the best regression function under the squared-error loss

may arise from diverse scientific communities. To avoid confusion, the word algorithm can be used exchangeably with function, model, or rule. Using the dataset **tr** for learning, training, or fitting, means replacing, or estimating, the algorithm's unknown parameters with appropriate values, as will be explained throughout the chapter. Therefore, at the end of this learning process, the final algorithm, function, model, or rule, is called learned, trained, or fitted.

1.3 Roadmap

The remainder of this chapter is structured as follows. Section 2 introduces the statistical decision theory, which constitutes the foundation of ML. The chapter explains how the ideal (the best performing) ML algorithm can be constructed, either for regression or classification, if we know the probability distribution of the data. Section 3 introduces some important parametric models for both regression and classification, and how they are constructed. Section 4 introduces the nonparametric and smoothing models, and explains the connection to neural network. These three sections will follow [20], an excellent comprehensive source for regression and classification methods with practical approaches and illustrative examples. Section 5 introduces mathematical optimization and how it is strongly connected to the construction of ML algorithms. This section will follow [8]. Section 6 discusses, in more detail, the performance of classification rules. It provides the link between the present and the next chapter. Section 7 concludes the chapter and provides a general advice for practitioners.

2 Statistical Decision Theory

This section provides an introduction to statistical decision theory, which serves as the foundation of ML. If a random vector X and a random variable Y have a joint probability density function (PDF) $f_{X,Y}(x, y)$ the problem is defined as follows: how to predict the variable Y from an observed value for the variable X. In this section we assume having a full knowledge of the joint density $f_{X,Y}$; therefore, there is no learning yet (Definition 1). The prediction function $\eta(X)$ is required to have minimum average prediction error. The prediction error should be defined in terms of some loss function $L(Y, \eta(X))$ that penalizes for any deviation in the predicted value of the response from the correct value. Define the predicted value by:

$$\hat{Y} = \eta(X). \tag{1}$$

The risk of this prediction function is defined by the average loss, according to the defined loss function:

$$\mathrm{R}(\eta) = \mathrm{E}\, L(Y, \hat{Y}). \tag{2}$$

2.1 *Regression*

Suppose that the response Y is a quantitative variable. This is the starting point of the statistical branch of regression, where (1) is the regression function. A form should be assumed for the loss function. A mathematically convenient and widely used form is the squared-error loss function:

$$L\left(Y, \eta\left(X\right)\right) = \left(Y - \eta\left(X\right)\right)^2. \tag{3}$$

In this case (2) becomes:

$$\mathrm{R}(\eta) = \int \left(Y - \eta(X)\right)^2 dF_{X,Y}(X, Y) \tag{4a}$$

$$= \mathrm{E}_X\, \mathrm{E}_{Y|X}\left[\left(Y - \eta(X)\right)^2 | X\right]. \tag{4b}$$

Hence, (4b) is minimized by minimizing the inner expectation over every possible value for the variable X; and the best regression function is then given by:

$$\eta^*(X) = \arg\min_{\eta(X)} \left[\mathrm{E}_{Y|X}\left[(Y - \eta(X))^2|X\right]\right] \tag{5a}$$

$$= \mathrm{E}_Y\left[Y|X\right] \tag{5b}$$

This means that if the joint distribution for the response and predictor is known the best regression function, in the sense of minimizing the risk, is the expectation of the response conditional on the predictor (Fig. 1). In that case the risk of regression in (4b) will be:

$$\mathrm{R}_{\min}(\eta) = \mathrm{R}(\eta^*) = \mathrm{E}_X \operatorname{Var}\left[Y|X\right]. \tag{6}$$

2.2 *Classification*

Recalling (2), and supposing that the response is a qualitative (or categorical) variable, give rise to the classification problem. Now the loss function cannot be the squared-error loss function defined in (3), because this has no meaning for categorical variables. Because Y may take now a qualitative value from a set of size K (Sect. 1), the loss function can be defined by the matrix

$$L(Y, \eta(X)) = \left(\left(c_{ij}\right)\right), \quad 1 < i, j < K, \tag{7}$$

where the non-negative element c_{ij} is the cost, the penalty, or the price, paid for classifying an observation as y_j when it belongs to y_i. Under this assumption, the risk defined by (2) can be rewritten for the categorical variables to be:

$$\mathrm{R}(\eta) = \mathrm{E}_X\, \mathrm{E}_{Y|X}\, L\left(Y, \eta(X)\right) \tag{8a}$$

$$= \mathrm{E}_X \sum_{i=1}^{K} c_{ij} \Pr\left[Y = y_i|X\right], \tag{8b}$$

where $\Pr\left[Y|X\right]$ is the probability mass function for Y conditional on X. Then, the conditional risk for the decision y_j,

$$\mathrm{R}(j, \eta) = \sum_{i=1}^{K} c_{ij} \Pr\left[Y = y_i|X\right], \tag{9}$$

is the expected loss when classifying an observation as belonging to y_j, where the expectation is taken over all the possible values of the response. Again, (8b) can be minimized by minimizing the inner expectation to give:

$$\eta^*(X) = \arg\min_j \left[\sum_{i=1}^{K} c_{ij} \Pr\left[Y = y_i|X\right]\right]. \tag{10}$$

Expressing the conditional probability of the response in terms of Bayes law, and substituting in (10) gives:

$$\eta^*(X) = \arg\min_j \left[\sum_{i=1}^{K} c_{ij} f_X(X|Y = y_i) \Pr[y_i] \right]. \tag{11}$$

The probability $\Pr[y_i]$ is the prior probability for y_i, while $\Pr[y_i|X]$ is the posterior probability, i.e., the probability that the observed case belongs to y_i, given the value of X. This is what is called Bayes classification, Bayes decision rule, or alternatively, the Bayes classifier.

Some special cases here may be of interest. The first case is when equal costs are assigned to all misclassifications and there is no cost for correct classification, i.e., $c_{11} = c_{22} = 0$ and $c_{12} = c_{21} = 1$, which is called the 0–1 cost, or loss function. This reduces (10) to:

$$\eta^*(X) = \arg\min_j \left[1 - \Pr[Y = y_j|X]\right] \tag{12a}$$

$$= \arg\max_j \left[\Pr[Y = y_j|X]\right]. \tag{12b}$$

The rule thus is to classify the observed case to the class having maximum posterior probability, which is very intuitive.

Another special case of great interest is binary classification, i.e., the case of $K = 2$. In this case (10) reduces to:

$$\frac{\Pr[y_1|X]}{\Pr[y_2|X]} \underset{y_2}{\overset{y_1}{\gtrless}} \frac{(c_{22} - c_{21})}{(c_{11} - c_{12})}. \tag{13}$$

Alternatively, this can be expressed as:

$$\frac{f_X(X = x|y_1)}{f_X(X = x|y_2)} \underset{y_2}{\overset{y_1}{\gtrless}} \frac{\Pr[y_2](c_{22} - c_{21})}{\Pr[y_1](c_{11} - c_{12})}. \tag{14}$$

The decision taken in (10) has the minimum risk, which can be calculated by substituting back in (8b) to give:

$$\mathrm{R}_{\min}(\eta) = \sum_{i=1}^{K} \int_X c_{ij} \Pr[y_i] \, dF_X(X|y_i), \tag{15}$$

where $j = \eta(X)$, which is the class decision prediction.

For the case where $K = 2$ and $c_{ii} = 0,\ i = 1, 2$, Eq. (15) reduces further to:

$$\mathrm{R}_{\min}(\eta) = c_{12} \Pr[y_1] \int\limits_{R_2} dF_X(X|y_1) + c_{21} \Pr[y_2] \int\limits_{R_1} dF_X(X|y_2), \tag{16}$$

where each of R_1 and R_2 is the predictor hyperspace over which the optimum decision (13) predicts as class 1 or class 2, respectively. Later, the response variable Y may be referred to Ω in case of classification; and to follow the notation of Sect. 1, the response of an observation is assigned a value ω_i, $i = 1, \ldots, K$, to express a certain class.

Example 1 Figure 2 illustrates an example of a binary classification problem, where each class has a two dimensional predictor, with a binormal distribution, with two different mean vectors μ_1, μ_2, and two different covariance matrices Σ_1, Σ_2. The best decision surface appears as the intersection of the two PDFs (left). The observations sampled from these two classes, along with this best decision surface, are drawn in the 2D space of the predictors (right). It is interesting, and may be counter-intuitive for some practitioners, to know that although the two distributions are normally distributed, the likelihood ratio (14) is not necessarily normally distributed [28]. For an early development of the theory of binary classification under the multinormal assumption of the class distribution, [17] is an indispensable resource. □

2.3 Where Is Learning?

To recap, this section emphasized the fact that there is no distinction between regression and classification from the conceptual point of view. Each minimizes the risk of predicting the response variable for an observation, i.e., a sample case with known predictor(s). If the joint PDF for the response and predictors is known, it is just a matter of direct substitution in the above results, which produces the best regression or classification function that minimizes the risk. If the joint distribution is known but its parameters are not known, e.g., multinormal distribution with unknown mean vector and covariance matrix, a learning process in this case is nothing but estimating those parameters from the dataset **tr** by well known methods of statistical inference. However, if the joint distribution is unknown, this gives rise to two different branches of prediction: (1) parametric regression (or classification), where the regression or classification function is modeled and a training sample is used to build that model, (2) and nonparametric regression (or classification), where no particular parametric model is assumed. Subsequent sections in this chapter briefly review some of these techniques, and explain the interesting connections among them.

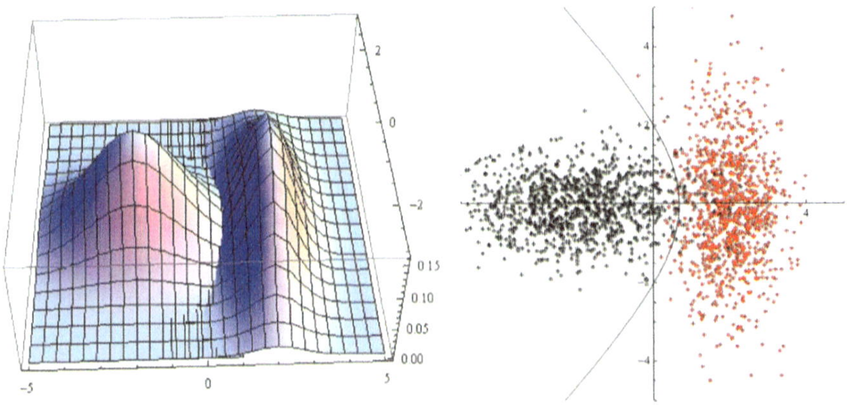

Fig. 2 The best decision surface of a binary classification problem with binormal features: the two PDFs, with their intersection that shows the best decision surface (left); and how the decision surface looks in the 2D feature space, along with observations drawn from the two classes (right)

3 Parametric Regression and Classification

The prediction method introduced in Sect. 2 assumes, as indicated, that the joint PDF of the response and the predictor is known. If such knowledge does not exist all the methods revolve around modeling the regression function (1) in the case of regression or the posterior probabilities in (10) in the case of classification.

3.1 *Linear Models (LM)*

In LM theory, it is assumed that Y is in the form:

$$Y = \mathrm{E}\,Y + e \tag{17a}$$

$$= \alpha + X'\beta + e, \tag{17b}$$

where the randomness of Y comes only from e, the conditional expectation of Y is linear in the predictors X, and the random error component e has a zero mean and a constant variance with X. The regression function (1) is then written as:

$$\eta(X) = \alpha + X'\beta. \tag{18}$$

More generally, still a LM, it can be rewritten as:

$$\eta(X) = X'_{new}\beta, \tag{19a}$$

$$X'_{new} = (f_1(X), \ldots, f_d(X)), \tag{19b}$$

where the predictor X is replaced by a new d-dimensional vector, X_{new}, whose elements are scalar functions of the original random vector X. The intercept α in (18) may be absorbed in terms of (19a) by setting $f_1(X) = 1$. Equation (19a) can be seen as equivalent to (18), where X has been transformed to X_{new}, which became the new predictor, on which Y will be regressed.

Now β must be estimated, and this point estimation is done for some observed values of the predictor; this is merely the learning process of the LM. Writing the equations for n observed values gives:

$$\mathbf{y} = \mathbf{X}\beta + \mathbf{e}. \tag{20}$$

Eq. (20) can be solved for β to give the least sum-of-squares for the components of error vector $\mathbf{e}$, which is quite known as the least-squares (LS) problem (Sect. 5). Said differently, it can be solved to minimize the residual sum-of-squares (RSS) between the predicted and the true response:

$$\begin{aligned} \text{RSS} &= \mathbf{e}'\mathbf{e} && \text{(21a)}\\ &= (\mathbf{y} - \mathbf{X}\beta)'(\mathbf{y} - \mathbf{X}\beta) && \text{(21b)}\\ &= \sum_i (y_i - x_i'\beta)^2, && \text{(21c)} \end{aligned}$$

to give:

$$\widehat{\beta} = \left(\mathbf{X}'\mathbf{X}\right)^{-1} \mathbf{X}'\mathbf{y}. \tag{22}$$

Then the prediction $\widehat{Y}$ of Y is done by estimating its expectation, which is given by:

$$\widehat{Y} = \widehat{\eta}(X) = \widehat{\mathrm{E}\,Y} = X'\widehat{\beta}. \tag{23}$$

For short notation we always write $\widehat{Y}$ instead of $\widehat{\mathrm{E}\,[Y]}$. The rational behind minimizing the RSS is that RSS$/n$ is a good estimate of the mean squared error (MSE), or the expected squared-loss $\mathrm{E}(Y - X'\beta)^2$. In addition, the latter is differentiable, which leads to the closed-form solution (22).

Nothing up to this point involves statistical inference. This is just fitting a mathematical model using the squared-error loss function. Statistical inference starts when considering the random error vector $\mathbf{e}$ and the effect of that on the confidence interval for $\hat{\beta}$, and the confidence in predicted values of the response for particular predictor variable, or any other needed inference. All of these important questions are answered by the theory of LMs. Bowerman and O'Connell [7] is a very good reference for an applied approach to LMs, without any mathematical proofs. For a theoretical approach and derivations, the reader is referred to [12, 18, 23].

It is remarkable that if the joint distribution of the response and the predictor is multinormal, the LM assumption (17b) is an exact expression of the random variable Y. This result arises from the fact that the conditional expectation of the multinormal distribution is linear in the conditional variable. That is, by assuming the joint PDF is multinormal with mean vector μ and covariance matrix Σ, and given by:

$$\begin{pmatrix} Y \\ X \end{pmatrix} \sim N(\mu, \Sigma), \quad \mu = \begin{pmatrix} \mu_Y \\ \mu_X \end{pmatrix}, \quad \Sigma = \begin{pmatrix} \Sigma_{11} & \Sigma_{12} \\ \Sigma_{21} & \Sigma_{22} \end{pmatrix}, \tag{24}$$

then the conditional expectation of Y on X is given by:

$$\mathrm{E}[Y | X = x] = \mu_Y + \Sigma_{12}\Sigma_{22}^{-1}(x - \mu_X). \tag{25}$$

For more details on the multinormal properties see [1].

In the case of classification, the classes are categorical variables but a dummy variable can be used as coding for the class labels. Then a linear regression is carried out for this dummy variable on the predictors. A drawback of this approach is what is called class masking, i.e., if more than two classes are used, one or more can be masked by others and they may not be assigned to any of the observations in prediction. For a clear example of masking see [20, Sect. 4.2].

3.2 Generalized Linear Models (GLM)

In a LM, the response variable is directly related to the regression function by a linear expression of the form (17b). In many cases a model can be improved by indirectly relating the response to the predictor through a LM—some times it is necessary, as well, for the classification problem, as will be shown. This is done through a transformation or a *link* function g, by assuming:

$$g(\mathrm{E}\,Y) = X'\beta. \tag{26}$$

Now it is the transformed expectation that is modeled linearly. Hence, LMs are merely a special case of the GLM when the link function is the identity function $g(\mathrm{E}\,Y) = \mathrm{E}\,Y$.

A very useful link function is the *logit* function defined by:

$$g(\mu) = \log \frac{\mu}{1 - \mu}, \quad 0 < \mu < 1. \tag{27}$$

Through this function the regression function is modeled in terms of the predictor as:

$$\mathrm{E}[Y] = \frac{\exp(X'\beta)}{1 + \exp(X'\beta)}, \tag{28}$$

which is known as logistic regression (LR). Equation (28) implies a constraint on the response Y, i.e., it must satisfy $0 < \mathrm{E}[Y] < 1$, a feature that makes LR an ideal approach for modeling the posterior probabilities in (10) for the classification problem. Equation (27) models the two-class problem, i.e., binary classification, by considering the new responses Y_1 and Y_2 to be defined in terms of the old responses ω_1 and ω_2, the classes, as:

$$Y_1 = \Pr\left[\omega_1|X\right], \tag{29a}$$

$$Y_2 = \Pr\left[\omega_2|X\right] = 1 - \Pr\left[\omega_1|X\right]. \tag{29b}$$

The general case of the K-class problem can be modeled using $K - 1$ equations, because of the constraint $\sum_k \Pr\left[\omega_k|X\right] = 1$, as:

$$\log \frac{\Pr\left[\omega_k|X = x\right]}{\Pr\left[\omega_K|X = x\right]} = x'\beta_k, \quad k = 1, \ldots, K - 1. \tag{30}$$

Alternatively, (30) can be rewritten as:

$$\Pr\left[\omega_k|X = x\right] = \frac{\exp\left(x'\beta_k\right)}{1 + \sum_{k'=1}^{K-1} \exp\left(x'\beta_{k'}\right)}, \quad 1 \leq k \leq K - 1, \tag{31}$$

$$\Pr\left[\omega_K|X = x\right] = \frac{1}{1 + \sum_{k'=1}^{K-1} \exp\left(x'\beta_{k'}\right)}. \tag{32}$$

The question now is how to estimate $\beta_k \; \forall \; k$. The multinomial distribution for modeling observations is appropriate here. For illustration, consider the case of binary classification; the log-likelihood for the n-observations can then be written as:

$$l(\beta) = \sum_{i=1}^{n} \left[y_i \log \Pr[\omega_1|X_i, \beta] + (1 - y_i) \log(1 - \Pr[\omega_1|X_i, \beta])\right] \tag{33a}$$

$$= \sum_{i=1}^{n} \left[y_i x_i'\beta - \log(1 + e^{x_i'\beta})\right]. \tag{33b}$$

To maximize this likelihood, the first derivative is set to zero to obtain:

$$\frac{\partial l(\beta)}{\partial \beta} = \sum_{i=1}^{n} x_i \left(y_i - \frac{e^{x_i'\beta}}{1 + e^{x_i'\beta}}\right) \stackrel{set}{=} 0. \tag{34}$$

This is a set of p, or d, nonlinear equations, because the vector X can be either the original predictor $(x_1, \ldots, x_p)'$ or any transformation $(f_1(X), \ldots, f_d(X))'$ as in (19b). These equations can be solved by iterative numerical methods like the Newton-Raphson algorithm. Finding the optimal values of these parameters is one of the optimization problems (Sect. 5), whose solution exists in many software packages. For more details with numerical examples see [20, Sect. 4.4] or [9, Sect. 12.3].

It can be noted that (33a) is valid under the assumption of the following general distribution:

$$f(X) = \phi(\theta_i, \gamma)h(X, \gamma)\exp(\theta_i' X), \tag{35}$$

with probability $p_i, i = 1, 2, p_1 + p_2 = 1$, which is the exponential family. So LR is no longer an approximation for the posterior class probability if the distribution belongs to the exponential family. For insightful comparison between LR and the Bayes classifier under the multinormal assumption see [14].

It is very important to mention that LR, and all subsequent classification methods, assume equal a priori probabilities. Then the ratio between the posterior probabilities will be the same as the ratio between the densities that appear in (11). Hence, the estimated posterior probabilities from any classification method are used in (11) as if they are the estimated densities.

3.3 Nonlinear Models

The link function in the GLM is modeled linearly in the predictors (26). Consequently, the response variable is modeled as a nonlinear function. In contrast to the LMs described in Sect. 3.1, in nonlinear models the response can be modeled nonlinearly right from the beginning, without the need for a link function.

4 Nonparametric Regression and Classification

In contrast to parametric regression, the regression function (1) is not modeled parametrically; i.e., there is no particular parametric form to be imposed on the function. Nonparametric regression is a versatile and flexible method of exploring the relationship of two variables. It may appear that this technique is more efficient than the LMs, but this is not the case. LMs and nonparametric models can be thought of as two different techniques in the analyst's toolbox. If there is an a priori reason to believe that the data follow a parametric form, then LMs or parametric regression in general may provide an argument for an optimal choice. If there is no prior knowledge about the parametric form the data may follow, or no prior information about the physical phenomenon that generated the data, there may be no choice other than nonparametric regression. There are many nonparametric techniques proposed in the statistical literature. What was said above, when comparing parametric and nonparametric methods, can also be said when comparing nonparametric methods to each other. None can be preferred across all situations (Sect. 7).

4.1 *Smoothing Techniques*

Smoothing is a tool for summarizing, in a nonparametric way, a trend between a response and a predictor such that the resulting relationship is less variable than the original response, hence the name smoothing. When the predictor is uni-dimensional, the smoothing is called scatter-plot smoothing. In this section, some methods used in scatter-plot smoothing are considered. These smoothing methods do not succeed in higher dimensionality. This is one bad aspect of what is called the curse of dimensionality (Sect. 6.5).

4.1.1 *K*-Nearest Neighbor (KNN)

The regression function (1) is estimated in the KNN approach by:

$$\eta(x) = \frac{1}{n}\sum_{i=1}^{n} y_i W_i(x), \tag{36}$$

$$W_i(x) = \begin{cases} n/K & i \in \mathscr{I}_x = \{i : x_i \in N_K(x)\} \\ 0 & otherwise \end{cases}, \tag{37}$$

where $N_K(x)$ is the set consisting of the nearest K points to the point x. In words, this technique approximates the conditional mean, i.e., the regression function that gives minimum risk, by local averaging the response Y.

In the case of classification, the posterior probability is estimated by:

$$\Pr\left[\omega_j | x\right] = \frac{1}{n}\sum_{i=1}^{n} I_{\omega_i=\omega_j} W_i(x), \tag{38}$$

and I is the indicator function defined by:

$$I_{cond} = \begin{cases} 1 & cond = True \\ 0 & cond = False \end{cases}. \tag{39a}$$

That is, replacing the continuous response in (36) by an indicator function for each class given each observation. So, the posterior probability is approximated by a frequency of occurrence in a K-point neighborhood.

A single-nearest-neighbor method (1-NN) is a special case of the KNN method, where $K = 1$. It can be thought of as narrowing the window W on which regression are carried out. In effect, this makes the regression function or the classifier more complex because it is trying to estimate the distribution at each point, which results in decreasing the bias and increasing the variance (Sect. 6.4).

4.1.2 Kernel Smoothing

In this approach, a kernel smoothing function κ is assumed. This means that a weighting and convolution (or mathematical smoothing) is carried out for the points in the neighborhood of the predicted point according to the chosen kernel function. Formally this is expressed as:

$$\eta(x) = \sum_{i=1}^{n} y_i \kappa \left(\frac{x - x_i}{h_x} \right) \Big/ \sum_{i'=1}^{n} \kappa \left(\frac{x - x_{i'}}{h_x} \right). \tag{40}$$

Choosing the bandwidth h_x of the kernel function is not an easy task. Usually, it is done numerically by cross validation (as explained in the next chapter). It is worth remarking that KNN smoothing is nothing but a kernel smoothing for which the kernel function is an unsymmetrical flat window spanning the range of the K-nearest neighbors of the point x. The kernel (40) is called Nadaraya-Watson kernel. Historically, and interestingly, [22] first introduced the window method density function estimation; his work was pioneered later by [21, 26] in regression.

4.2 *Additive Models (AM)*

Recalling (19), and noticing that the function $f_i(X)$ is a scalar parametric function of the whole predictor, show that LMs are parametric AMs. By dropping the parametric assumption and letting each scalar function be a function of just one element of the predictor, i.e., X_i, allows defining a new nonparametric regression method, namely AMs, as:

$$\eta(x) = \alpha + \sum_{i=1}^{p} f_i(X_i), \tag{41}$$

where the predictor is of p dimensions. The response variable itself, Y, is modeled as in (17a) by assuming zero mean and constant variance for the random component e. Then, $f_i(X_i)$ is fit by any smoothing method defined in Sect. 4.1. Every function $f_i(X_i)$ fits the value of the response minus the contribution of the other $p - 1$ functions from the previous iteration. This is called the back-fitting algorithm [19, Sect. 4.3]

4.3 *Generalized Additive Models (GAM)*

GAMs can be developed in a way analogous to how GLMs were developed above, i.e., by working with a transformation of the response variable, hence the name generalized additive models. Equation (41) describes the regression function as an AM; alternatively it can be described through another link function:

$$g(\eta(x)) = \alpha + \sum_{i=1}^{p} f_i(X_i). \tag{42}$$

Again, if a *logit* function is used the model can be used for classification exactly as was done in the case of GLMs. Rewriting the score Eq. (34) for the GAM, using the posterior probabilities as the response variable, produces the nonparametric classification method using the GAM. Details of fitting the model can be found in [19, Sect. 4.5 and Chap. 6].

4.4 Projection Pursuit Regression (PPR)

PPR, introduced by [16], is a direct attack on the dimensionality problem, since it considers the regression terms as a summation of terms, each of which is a function of a projection of the whole predictor onto a direction (specified by some unit vector). Formally it is expressed as:

$$\eta(x) = \sum_{i=1} g_i(\alpha_i' x). \tag{43}$$

The function g_i, for every selection of the direction α_i, is to be fit by a smoother in the new single variable $\alpha_i' x$. It should be noted that (43) assumes that the function $g_i(\alpha_i' X)$, named the *ridge function*, is constant along any direction perpendicular to α_i. Fitting the model is done by iteratively finding the best directions α_i's that minimize(s) the RSS, hence the name pursuit. Details of fitting the model and finding the best projection directions can be found in [16, 20].

In (43), by deliberately setting each unit vector α_i to have zero components except $\alpha_{ii} = 1$, reduces the PPR to AM. Moreover, and interestingly as well, introducing the *logit* link function to the regression function $\eta(x)$ in (43) suits the classification problem exactly as was done in the GAM. This turns out to be exactly the same as the single-hidden-layer NN, as will be presented in the next section.

4.5 Neural Networks (NN)

The field of NN has been evolving, since its start in the engineering community around 1950s, until we reached now the era of deep neural networks (DNN). A single-hidden-layer NN can be considered as a process for modeling the output in terms of a linear combination of the inputs. The set of p input features, i.e., the predictor components $X_1, \ldots, X_p$, are in turn weighted linearly to form a new set of M arguments, $Z_1, \ldots, Z_M$, that go through the sigmoid function σ, which can have different values of steepness, or learning rate. Figure 3 illustrates a single-hidden-layer NN with its architecture (left), and a plot of its sigmoid function with

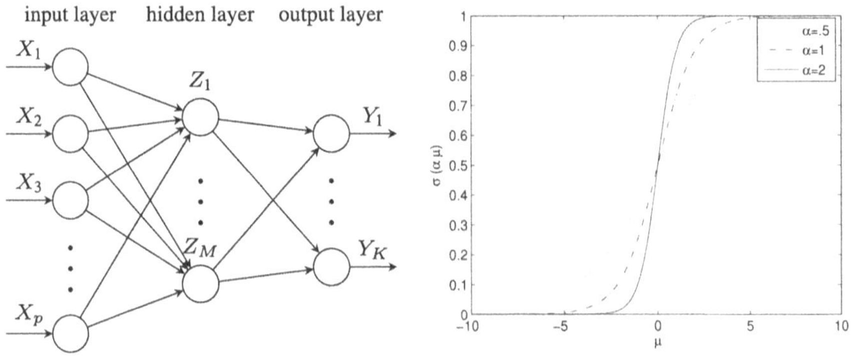

Fig. 3 A single-hidden-layer NN. The architecture that reflects Eq. (44) (left), and the sigmoid function with different learning rates a (right)

different learning rates (right). The output of the sigmoid function accounts for a hidden layer consisting of M intermediate values. Then these M hidden values are in turn weighted linearly to form a new set of K arguments that go through the final output functions, whose output is the response variables $Y_1, \ldots, Y_K$. This can be expressed mathematically in the form:

$$Z_m = \sigma(\alpha_{om} + \alpha_m' X), \quad m = 1, \ldots, M, \tag{44a}$$

$$\sigma(\mu) = \frac{1}{1 + e^{-\mu}}, \tag{44b}$$

$$Y_k = f_k\left(\beta_{0k} + \sum_{m=1}^{M} \beta_{mk} Z_m\right), \quad k = 1, \ldots, K. \tag{44c}$$

Equation (44c) shows that if the function f is chosen to be the identity function, i.e., $f(\mu) = \mu$, the NN is simply a special case of the PPR method defined in (43), where the sigmoid function has been explicitly imposed on the model rather than being developed by any smoothing mechanism as in PPR. This is what is done when the output of the network is quantitative. When it is categorical, i.e., the case of classification, the function f can be simply modeled as:

$$f_k(\mu_k) = e^{\mu_k} \Big/ \sum_{k'=1}^{K} e^{\mu_{k'}}. \tag{45}$$

In this case each output node models the posterior probability $\Pr[\omega_k | X]$, which is exactly what is done by the LR link function defined in (27). Again, the model will be an extension to the GAM as defined at the end of Sect. 4.4. Although Eq. (44) are indeed parametric, we list NN in this section for the strong connection to the AM, GAM, and PPR that were just explained. Excellent references for the early basics and foundations of NN are [5, 24]. We conclude this section by quoting the following statement from [20]:

> There has been a great deal of hype surrounding neural networks, making them seem magical and mysterious. As we make clear in this section, they are just nonlinear statistical models, much like the projection pursuit regression model discussed above.

5 Optimization

Optimization serves an amazing variety of practical problems: e.g., optimizing power consumption in electrical stations, optimizing overall budget in project management, and most importantly to us in this chapter optimizing ML algorithms to provide the best performance. In this section, we will provide a very basic introduction to optimization and its strong connection to the construction of ML algorithms.

5.1 *Introduction*

The mathematical optimization problem (MOP) is an abstraction of how to make the "best" possible choice of some vector β under some *constraints*. These constraints represent a set of trim requirements, or specifications, that limits the possible choices of this vector. The *objective function* of this problem represents the *cost*, or *loss*, to minimize, or the *utility* to maximize, for each vector β, and this what makes that value of β the "best" possible choice. This is formalized in the following definition.

Definition 3 *A mathematical optimization problem* has the form:

$$\begin{array}{ll} \underset{\beta}{\text{minimize}} & f_0(\beta) \\ \text{subject to:} & f_i(\beta) \leq 0, \quad i = 1, \ldots, m, \\ & h_i(\beta) = 0, \quad i = 1, \ldots, l, \end{array}$$

where

$$\begin{array}{lr} \beta = (\beta_1, \ldots, \beta_p) \in \mathfrak{R}^p, & \text{(optimization variable)} \\ f_0 : \mathfrak{R}^p \mapsto \mathfrak{R}, & \text{(objective (cost) function)} \\ f_i : \mathfrak{R}^p \mapsto \mathfrak{R}, & \text{(inequality constraints (functions))} \\ h_i : \mathfrak{R}^p \mapsto \mathfrak{R}, & \text{(equality constraints (functions))} \\ \mathscr{D} : \bigcap_{i=1}^m \mathbf{dom} f_i \cap \bigcap_{i=1}^l \mathbf{dom} h_i & \text{(domain of constraints: feasible set)} \\ = \{\beta \mid \beta \in \mathfrak{R}^p \wedge f_i(\beta) \leq 0 \wedge h_i(\beta) = 0\} & \\ \beta^* : \{\beta \mid \beta \in \mathscr{D} \wedge f_0(\beta) \leq f_0(\alpha) \ \forall \alpha \in \mathscr{D}\}, & \text{(solution)} \end{array}$$

where the solution β^* is called the optimizer (or minimizer). □

The problem aims at minimizing a mathematical function, under some constraints. From Definition 3, it is clear that minimizing f_0 is the same problem as maximizing

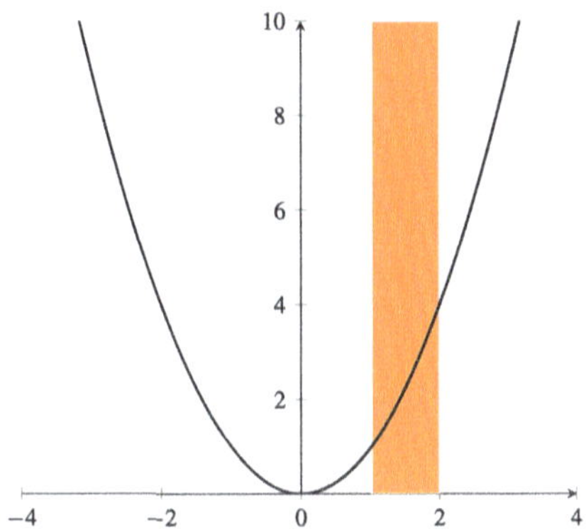

Fig. 4 An objective function in a single dimension, with a constraint $1 \leq \beta \leq 2$ (the colored region). The minimizer $\beta^* = 1$, under this constraint, is different from the global minimizer $\beta^* = 0$

$-f_0$; the constraints $f_i \leq 0$ are equivalent to $-f_i \geq 0$; the constraints $f_i \leq 0$ are equivalent to $f_i \leq b_i$, where b_i can be simply absorbed into f_i; and, finally, $m = l = 0$ is the case of unconstrained problem with global minimization.

Example 2 The is a very basic example of an MOP in a single dimension, with a single constraint:

$$\begin{aligned} &\underset{\beta}{\text{minimize}} && f_0(\beta) = \beta^2 \\ &\text{subject to:} && \beta \leq 2, \\ & && 1 \leq \beta. \end{aligned}$$

It is clear that the minimizer is $\beta^* = 1$; however, the minimizer for the unconstrained problem is $\beta^* = 0$ (Fig. 4). □

Example 3 [11, Example 20.1, p. 454]: This example shows how the MOP may not be as simple as finding the derivatives:

$$\begin{aligned} &\underset{\beta}{\text{minimize}} && f_0(\beta_1, \beta_2) = (\beta_1 - 1)^2 + \beta_2 - 2 \\ &\text{subject to:} && \beta_2 - \beta_1 = 1, \\ & && \beta_1 + \beta_2 \leq 2. \end{aligned}$$

This 2D objective function, along with the constraints, are illustrated in Fig. 5. It is obvious that the function has no global minimizer ($\partial f_0 / \partial \beta_2 = 1 \neq 0$). After setting the constraints, it is quite easy to see that $f_0|_{(\beta_2 - \beta_1 = 1)} = (\beta_1 - 1)^2 + (\beta_1 - 1)$ attains a minima at $\beta_1 = 1/2$, and hence, the minimizer is $\beta^* = (1/2, 3/2)'$. □

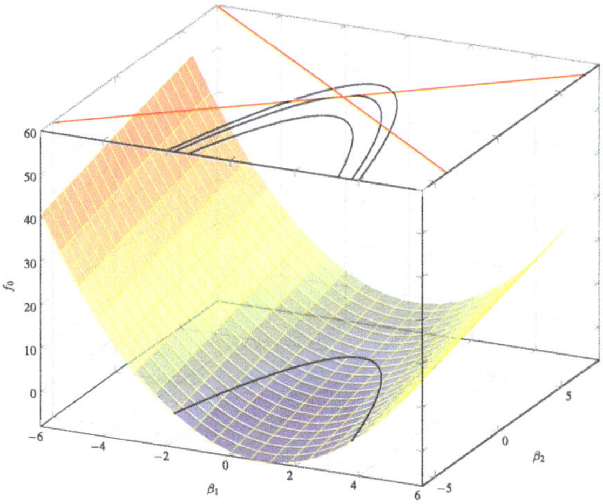

Fig. 5 A simple objective function in two dimensions (the colored surface, shown along with its contours drawn in black), with two constraints (the red lines). Although the surface has no global minimum, the constrained problem does have

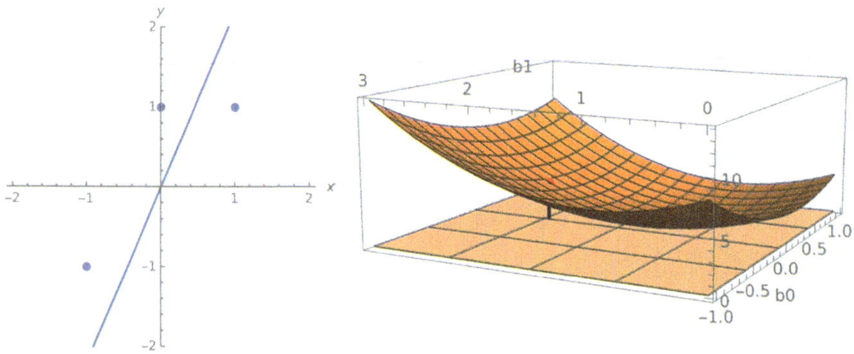

Fig. 6 ML and MOP. A training dataset of three observations, for a regression problem with a single feature, to be fitted by a linear model having only two parameters β_0 and β_1 (left). The RSS of this model is the objective function, of these two parameters, to be minimized (right). The red point on the surface is the value (not the minimum yet), at some initial values of β_0 and β_1 that corresponds to the intercept and slope of the line on the left

5.2 *Connection to Machine Learning*

As explained earlier in this chapter, all parametric ML algorithms, e.g., LM, LR, SVM, NN, DNN, etc., include parameters that need to be replaced by numerical values. This is performed with the help of a dataset that is called a training dataset. The following simple example illustrates the connection between ML and MOP, and relates both to the title of the present chapter.

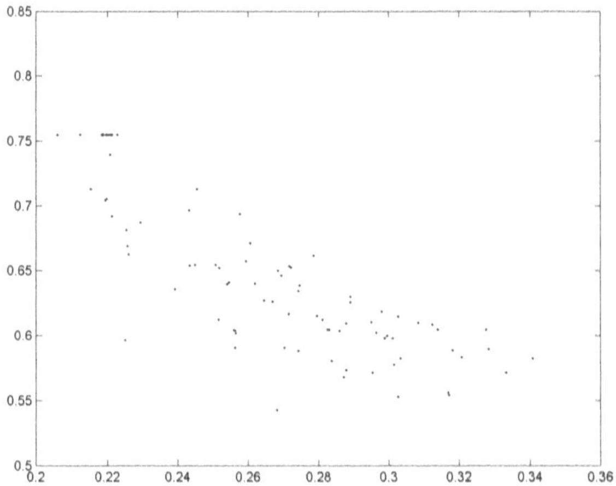

Fig. 7 100 pairs of true AUC versus true MSE. Each pair is obtained from training the same NN, to minimize the RSS, on a new training datasets, then testing on a very large testing dataset to mimic the population. Although there is an obvious trend of getting a high AUC with low MSE, it is not a guaranteed behaviour for each training dataset

Example 4 (Machine Learning: construction) Suppose that we have a strong belief that the best regression function for a particular problem is the LM $Y = \beta_0 + \beta_1 X$. Then, for a given training dataset $\mathbf{tr} : \{t_i = (x_i, y_i),\ i = 1, \ldots, n\}$, we need to minimize the RSS of this model on this dataset. This is a typical MOP, which can be formalized as:

$$\underset{\beta_0, \beta_1}{\text{minimize}} \sum_{i=1}^{n} (\beta_0 + \beta_1 x_i - y_i)^2, \quad (x_i, y_i) \in \mathbf{tr}.$$

Figure 6 illustrates a dataset of only three observations ($n = 3$), along with a straight line of initial values of the parameters $\beta_0,\ \beta_1$, all drawn in the feature space (left). The objective function to be minimized (the RSS) is drawn as a function of the two parameters β_0, β_1 (right). Each pair of values of $\beta_0,\ \beta_1$ results in a new line (fitted model) in the feature space, and a new point on the surface of the objective function in the parameter space. The solution of this MOP is the vector $\beta^* = (\beta_0, \beta_1)$ that minimizes the objective function, which fortunately for linear models has a closed-form solution given earlier in Eq. (22). □

Departing from the previous example, in the following few paragraphs we will emphasize important concepts. The example demonstrated the relationship between: (1) the ML model, along with the training dataset, in the feature space, and (2) the objective function, which should be minimized, in the parameter space. All other ML models, whether for regression or classification, have parameters that should be replaced, tuned, or estimated, to optimize (minimize or maximize) some objective function. Ideally, this objective function should be a good estimator for the same intended performance measure of the model, not for any other performance measure. However, some mathematical difficulties may preclude this ideal practice, as will be seen next.

In Example 4, the objective function to be minimized was the RSS, which minimizes RSS/n, an estimator of the MSE. However, in some circumstances it is very difficult mathematically to optimize the targeted performance measure. In such cases, another performance measure is optimized, because of the tractability of its mathematical formalization, hoping that the solution optimizes, as well, the targeted performance measure. Figure 7 illustrates this fact for a very simple four-neuron single-layer NN, trained on a simulated two-class univariate normal dataset. The NN is required to achieve a high AUC (a performance measure that will be explained in Sect. 6); however, because its estimator is non differentiable, and therefore is very hard to maximize using conventional mathematical approaches, the NN is trained to minimize the RSS, instead. For illustration, the NN is trained on 100 different training datasets, and tested after each training on a very large testing dataset to provide a good estimate of the true AUC and MSE. The figure shows the 100 pairs of values of these two performance measures, with a general trend of exhibiting a high AUC with a low MSE. However, some instances exhibited a low MSE (good performance) associated with a low AUC (bad performance).

Another important fact to emphasize is that all of the model's parameters that are estimated during the learning process are functions of thc training dataset. Hence, the following facts hold: these parameters are random variables, the model is a random model, the objective function is a random function, and the minimum value of this objective function, which is the model optimal performance, is a random variable (as will be detailed in Sect. 6); all will vary if the training dataset varies.

To recap, Example 4 demonstrated how LM, one of the ML models explained in this chapter, represents an MOP whose solution fortunately can be found in closed form. Other ML models belong to a class of MOP that is difficult to solve; DNN is an example. In the next section, we will provide a very short account of the taxonomy of the MOP to show its different types and the connection of each of these types to ML.

5.3 Types of MOP

According to the nature of the objective function and its constraints, the MOP can be classified into one of these nested classes:

$$\text{Linear} \subset \text{Quadratic} \subset \text{Convex} \subset \text{Nonlinear}.$$

For each of these classes, several questions arise: (1) is there a closed-form solution? (2) if not, is there a numerical solution? (3) if yes, is it guaranteed? (4) what are the ML models that belong to this class? In the following subsections, we discuss very briefly each of these classes, and provide some answers to these questions. It is important to emphasize that although these classes are mathematically nested—in the very strict sense that any linear is quadratic, any quadratic is convex, and any convex is nonlinear—the solution techniques for each class are quite different from others.

The solution techniques for these classes vary between: closed-form, numerical (e.g., Newton's methods, gradient descent, etc.), or even *intelligent*-based methods (e.g. genetic algorithms, particle swarm, etc.).

5.3.1 Linear Programming

Definition 4 A linear programming problem is an MOP with an objective and all constraints are linear:

$$\begin{aligned} \underset{\beta}{\text{minimize}} \quad & f_0(\beta) = c'\beta \\ \text{subject to:} \quad & a_i'\beta \leq b_i, \quad i = 1, \ldots, m, \\ & h_i'\beta = g_i, \quad i = 1, \ldots, l. \end{aligned}$$

□

Example 5 (Chebyshev minimization) The MOP:

$$\underset{\beta}{\text{minimize}}\, f_0(\beta) = \max_{i=1,\ldots,n} |y_i - x_i'\beta|,$$

can be understood in terms of ML terminology as minimizing the maximum possible error, measured in absolute deviance between the true response value y_i and the predicted value $x_i'\beta$. This should be contrasted with the least-squares MOP of the LM (as illustrated in Example 4 and will be more detailed in Sect. 5.3.2 below) in two important aspects: (1) the error is measured in terms of absolute deviance rather than squared difference. (2) the objective function here focuses only on the single observation that achieves the maximum error rather than summing over all observations. After little manipulations, the problem can be reduced, and found to be equivalent, to the following:

$$\begin{aligned} \underset{\beta}{\text{minimize}} \quad & t \\ \text{subject to:} \quad & x_i'\beta - t \leq y_i, \qquad i = 1, \ldots, n \\ & -x_i'\beta - t \leq -y_i, \qquad i = 1, \ldots, n, \end{aligned}$$

which is a typical linear programming problem, per Definition 4. □

In general, there is no closed-form solution to the linear programming problems. However, there exists a set of very robust, reliable, and computationally effective methods of numerical solutions: e.g., Dantzig's simplex and interior point that can solve problems with several thousands of variables.

5.3.2 Least-Squares (LS) Problems

Definition 5 A LS problem is an MOP with no constraints (i.e., $m = l = 0$), and an objective in the form:

$$\underset{\beta}{\text{minimize}}\, f_0(\beta) = \sum_{i=1}^{n} (x_i'\beta - y_i)^2 = \|\mathbf{X}_{n\times p}\beta_{p\times 1} - \mathbf{y}_{n\times 1}\|^2.$$

□

Example 6 (LM) The linear models for regression, discussed in Sect. 3.1 and Example 4, is a typical example for the LS problem, where the solution is given in the closed form by $\beta = \left(\mathbf{X}'\mathbf{X}\right)^{-1}\mathbf{X}'\mathbf{y}$. □

The algorithms for finding the matrix inversion and matrix multiplication in this closed-form solution exist in many scientific computing software, and this technology is quite mature even for thousands of variables.

There is a more elaborate version of the LS problem that is called weighted LS. This type of problem appears in ML, e.g., when more emphasis is required on some observations than others:

$$\underset{\beta}{\text{minimize}}\, f_0(\beta) = \sum_{i=1}^{n} w_i (x_i'\beta - y_i)^2,$$

or when it is required to penalize for using extra parameters to guard against overfitting (Sect. 6.4), an approach known as regularization:

$$\underset{\beta}{\text{minimize}}\, f_0(\beta) = \sum_{i=1}^{n} (x_i'\beta - y_i)^2 + \rho \sum_{j=1}^{p} \beta_j^2.$$

It is quite easy to show that both problems can be solved as LS problem, per Definition 5.

5.3.3 Convex Optimization

Definition 6 A convex optimization problem is an MOP with an objective and all constraints are convex:

$$\begin{aligned}
\underset{\beta}{\text{minimize}} \quad & f_0(\beta) \\
\text{subject to:} \quad & f_i(\beta) \le 0, && i = 1, \ldots, m, \\
& h_i(\beta) = 0, && i = 1, \ldots, l, \\
& f_i(a\alpha + b\beta) \le af_i(\alpha) + bf_i(\beta), a + b = 1, && 0 \le a, b, 0 \le i \le m, \\
& h_i(\beta) = c_i'\beta + d_i && 0 \le i \le p.
\end{aligned}$$

□

Example 7 (Lasso Regression) Similar to the penalized LS problem, lasso regression minimizes the RSS; however, it does so with an L_1 penalty rather than the L_2 of the LS problem. The problem is formalized as:

$$\begin{aligned} &\underset{\beta}{\text{minimize}} \sum_{i=1}^{n}(y_i - x_i'\beta)^2 \\ &\text{subject to: } \sum_{j=1}^{p} |\beta_j| \leq t, \end{aligned}$$

which can be shown to be equivalent to the MOP:

$$\underset{\beta}{\text{minimize}}\, f_0(\beta) = \sum_{i=1}^{n}(y_i - x_i'\beta)^2 + \rho \sum_{j=1}^{p} |\beta_j|,$$

The latter, in contrast to the LS penalization, has no closed-form solution because of the difficulty introduced by the non-differentiable term $|\beta_j|$. However, the numerical solution is quite feasible and reliable as all convex optimization problems are. □

The numerical solution of a convex optimization problem is well established through the methods of *interior point*, although no closed-form solution exists. Problems with thousands of variables can be solved robustly as in linear programming problems. In addition, many problems are initially formulated, then with some mathematical manipulation they can be transformed to a solvable convex problem.

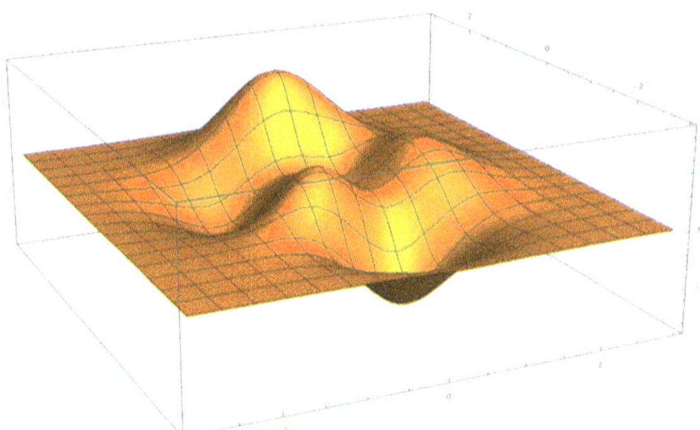

Fig. 8 A plot of a 3D function [11, Example 14.3, p. 290]: $f(\beta_0, \beta_1) = 3(1 - \beta_0)^2 e^{-\beta_0^2 - (\beta_1+1)^2} - 10e^{-\beta_0^2 - \beta_1^2}\left(-\beta_0^3 + \frac{\beta_0}{5} - \beta_1^5\right) - \frac{1}{3}e^{-(\beta_0+1)^2 - \beta_1^2}$ that shows several minima, maxima, and saddle points

5.3.4 Nonlinear Optimization

Definition 7 A nonlinear optimization problem is an MOP with objective and constraint functions are nonlinear □

Example 8 A nonlinear objective function, just in two dimensions, with several minima, maxima, and saddle points, is illustrated in Fig. 8. A NN, or in its more complex form, a DNN with several layers, can have hundreds of millions of parameters, not only two as illustrated in the figure! □

The nonlinear optimization problems can be very hard to solve, even for simple-looking problems in few parameters (variables). Several approaches exist for solving the problem; these approaches can be divided into two main categories: numberical methods and computational intelligence, as briefly explained in the following two paragraphs, respectively.

Finding the minima or the maxima of a function numerically is a well known topic in mathematics and numerical analysis. However, the challenge of nonlinear optimization, especially for problems like DNN, remains in the computational complexity that grows exponentially with the dimensions of the objective function, the matter that makes it almost impossible to find a global minimum. Alternatively, finding a local minimum is a practical compromise, although it does not guarantee converging to the global one. Local minimization starts at a point in the parameter space (usually is selected randomly, or by other criteria determined by the numerical algorithm) then the space is navigated, and guided by the multi-dimensional derivatives (with respect to the parameters) of the objective function. All the well known methods, starting form Newton's method to the most recent approaches used for DNN, e.g., stochastic gradient descent (SGD), belong to this category. It is obvious that the initial starting point in the parameter space heavily affects the convergence process and the final solution.

The term computational intelligence was first coined early by [2, 3]:

> A system is computationally intelligent when it: deals only with numerical (low-level) data, has a pattern recognition component, and does not use knowledge in the AI (Artificial Intelligence) sense; and additionally, when it (begins to) exhibit (i) computational adaptivity; (ii) computational fault tolerance; (iii) speed approaching human-like turnaround, and (iv) error rates that approximate human performance.

Since that time, the term computational intelligence (CI) has been accepted as a generic term to the field that combines NNs, fuzzy logic, and evolutionary algorithms [25, 31]. Later, the area of swarm detection was considered as a peer paradigm to the other three mentioned above [15].

6 Performance

From what has been early discussed at the beginning of this chapter, there is not any conceptual difference between regression and classification for the problem of supervised learning. Abstractly, both aim to achieve the minimum risk (2) under a certain loss function, for predicting a response, from a particular predictor. Although risk is a very obvious performance measure for assessing ML algorithms, we will elaborate in this section and show how we can depart and define other important performance measures, e.g., the individual error components, ROC, and AUC. It is must be noted that what will be defined in this section is the parametric form (also known as the true performance or the population performance), which can only be calculated if the posterior probabilities are known. On the contrary, if the posterior probabilities are not known all performance measures can be estimated from a given dataset, called the testing dataset, using appropriate estimators. If the testing dataset is infinitely large, i.e. testing on the population, the estimated performance will converge to the true performance. Performance estimation and different estimators are discussed in the next chapter.

6.1 Error Components

We will elaborate on the special case of binary classification, with no cost on correct classification ($c_{ii} = 0,\ i = 1, 2$), which is of great interest in many applications. In this case, the risk of each classifier is reduced to (16), which can be rewritten as:

$$\mathrm{R}_{\min} = c_{12} P_1 e_1 + c_{21} P_2 e_2, \tag{46}$$

where e_1 is the probability of classifying a case as belonging to class 2 when it belongs to class 1, and e_2 is vice versa.

In the feature space, the regions of classification have the dimensionality p, and it is very difficult to calculate the error components from multi-dimensional integration. It is easier to look at (14) as:

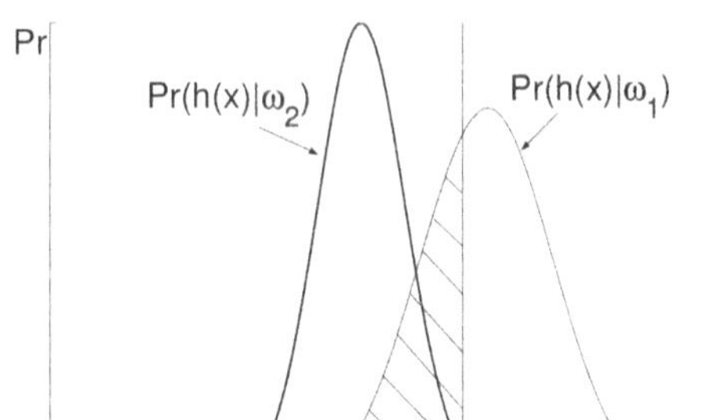

Fig. 9 The probability of LLR conditional on each class. The two components of error are indicated as the FPF and FNF

$$h(x) \underset{\omega_2}{\overset{\omega_1}{\gtrless}} th, \tag{47a}$$

$$h(x) = \log \frac{f_X(X = x|\omega_1)}{f_X(X = x|\omega_2)}, \tag{47b}$$

$$th = \log \frac{\Pr[\omega_1]c_{21}}{\Pr[\omega_2]c_{12}}, \tag{47c}$$

where the log is taken just as a convention to simplify the analysis for the case of multinormal distribution (because it has an exponent); however, it has no other significance. The function $h(X)$ is called the log-likelihood ratio (LLR), which is obviously a random variable, whose variability comes from the feature vector X. The LLR has a PDF conditional on each of the two classes, as indicated in Fig. 9; (it can be easily shown that the two curves in this figure cross at $h(X) = 0$, when the threshold is zero.)

In general, the two error components appearing in (46) can be rewritten, equivalently to their corresponding terms in Eq. (16), using the LLR in (47), as:

$$e_1 = \int_{-\infty}^{th} f_h\left(h(x)|\omega_1\right) dh(x), \tag{48a}$$

$$e_2 = \int_{th}^{\infty} f_h\left(h(x)|\omega_2\right) dh(x). \tag{48b}$$

Now, it is very important to realize the generality of this error equation and the two messages it conveys. (1) It expresses the two components of error for any classifier that produces an output, or a score, of $h(x)$ for a predictor $X = x$, even if it is not the best (Bayes') classifier. The only exception then would be that the score $h(X)$ is no longer the LLR that produces the minimum risk. (2) Whether $h(X)$ is the score of the Bayes' classifier or not, Eq. (48) says that at each threshold value th there is a pair of two components of error. Over the continuum of threshold values there is a continuum of these pairs, which define a new curve. This curve is called the ROC curve, a device that is much more rich for assessing classification rules than a single pair of errors, as will be explained next.

6.2 *Receiver Operating Characteristic (ROC) Curve*

Now, assume the classifier is trained under the condition of equal prevalence and cost, i.e., the threshold is zero. In other environments there will be different a priori probabilities yielding to different threshold values. The error is not a sufficient metric now, since it is a function of a single fixed threshold. A more general way to assess a classifier is provided by the ROC curve. This is a plot for the two components of error, e_1 and e_2, under different threshold values. It is conventional in many applications to

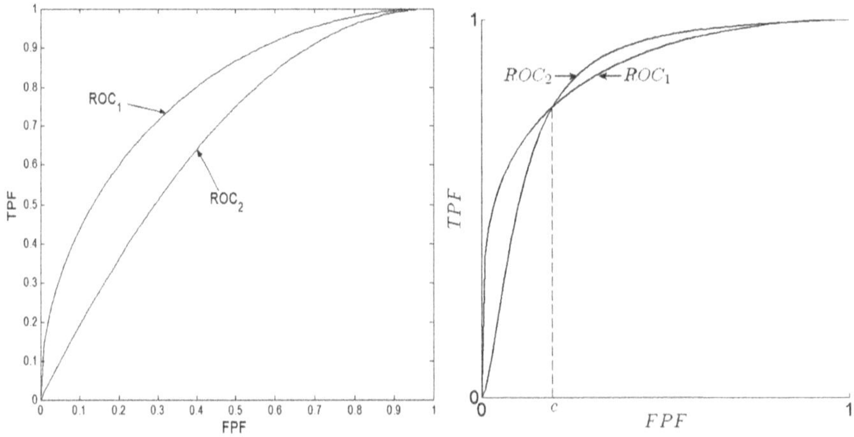

Fig. 10 ROC curves for two different competing classifiers. Left: ROC_1 is better than ROC_2, since for any error component value, the other component of classifier 1 is less than that of classifier 2. Right: ROC_1 is better than ROC_2 only in the range of FPF that is lower than the value c

refer to e_1 as the False Negative Fraction (FNF), and e_2 as the False Positive Fraction (FPF). This is because cases from the abnormal class typically are assigned higher classifier's scores than cases from the normal class, hence the names "positive" and "negative". For example, a network activity belonging to the abnormal class (the class of anomalous activities) whose classifier's score is less than the chosen threshold will be called "negative". This is obviously a false negative decision; hence the name FNF. The situation is reversed for the other error component.

Because the classification problem now can be seen, more generally, in terms of the classifier's output score rather than the hard binary decision, it is apparent that each of the two error components is an integral over a univariate PDF. Therefore, the resulting ROC is a monotonically non-decreasing function. A convention in many fields is to plot the true positive fraction (TPF), which is given by $\text{TPF} = 1 - \text{FNF}$, vs. the FPF. In that case, the farther apart the two distributions $f_h(h|\omega_i),\ i = 1, 2$ of the score function $h(X)$ from each other, the higher the ROC curve and the larger the area under the curve (AUC). Figure 10 (left) shows ROC curves for two different competing classifiers. The first classifier performs better because it has a lower value of e_2 at each value of e_1. Thus, the first classifier unambiguously separates the two classes better than the second one. Therefore, the AUC for the first classifier is larger than that for the second one. The AUC can be thought of as one summary performance measure for the ROC curve. Formally, the AUC is given by:

$$\text{AUC} = \int_0^1 \text{TPF}\ d(\text{FPF}). \tag{49}$$

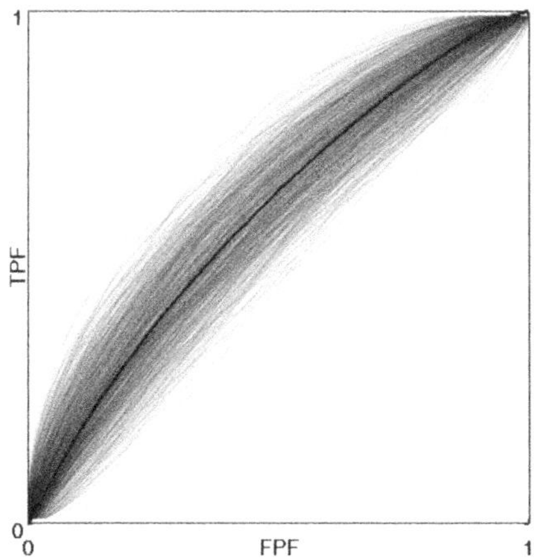

Fig. 11 A population of ROC curves of a classifier trained on several training datasets; for each training dataset an ROC is built. The mean ROC is shown in bold

And it can be shown that it is also given by:

$$\text{AUC} = \Pr\big[h(x)|\omega_2 < h(x)|\omega_1\big], \tag{50}$$

which expresses how the classifier scores for class ω_1 are stochastically larger than those of class ω_2, and hence more capable of the classification task.

If two ROC curves cross (Fig. 10, right), this means each classifier is better than the other only for a certain range of the threshold setting, and vice versa. In that case, some other performance measure can be used, such as the partial area under the ROC curve in a specified region [27].

6.3 The True Performance Is A Random Variable!

As was explained in Sect. 5.2, regardless of whether the ML task is regression or classification, the model, its parameters, and its performance, all are random variables, where the randomness comes from the training dataset. In addition, this variation depends on the complexity of the model, and its capacity to learn, relative to the training dataset size.

For instance, the output scoring function $h(X)$ of a particular classifier is indeed $h_{\mathbf{tr}}(X)$, which is subscripted to show the dependence on the training dataset; hence, the PDFs $f_{h_{\mathbf{tr}}}(h|\omega_i)$ and $\text{ROC}_{\mathbf{tr}}$, all should be subscripted as well. Therefore, there is a population of ROC curves that corresponds to the population of training datasets (Fig. 11). For more elaboration, consider the AUC as the performance measure of interest. Then, the fundamental quantities of interest are the following:

1. $\text{AUC}_{\mathbf{tr}}$: the true performance of the classifier, conditional on a particular training dataset **tr** of a specified size n but over the population of testing datasets (as if we trained on **tr** then tested on infinite number of observations),

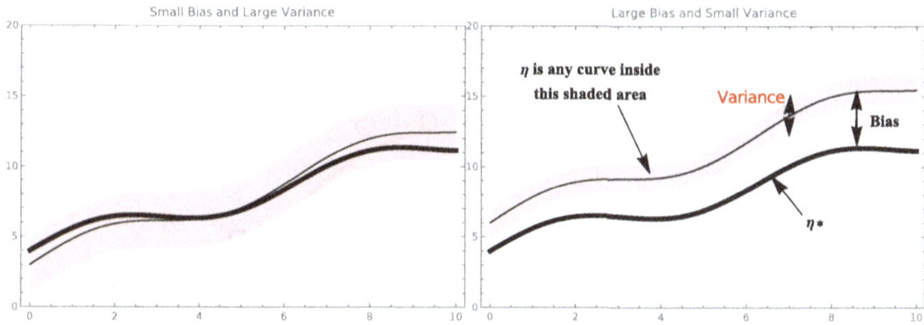

Fig. 12 Two regression functions with: low bias and high variance (left); high bias and low variance (right)

2. $\mathbf{E_{tr}}\,\mathrm{AUC}_{\mathbf{tr}}$: the expectation of the true performance over the population of training datasets of the same size n, and
3. $\mathrm{Var}_{\mathbf{tr}}\,\mathrm{AUC}_{\mathbf{tr}}$: the variance of the true performance over the population of training datasets of the same size n. This variance expresses how the classifier is sensitive to retraining, e.g. in the case of obtaining a new training dataset.

Any other performance measure, e.g., each of the error components, the risk, etc., is a r.v. as well, should be similarly subscripted **tr**, and has a mean and a variance as explained above. For more elaboration, we explain this crucial concept in Sect. 6.4, in a more mathematical detail, for the case of regression than classification, because it is more obvious and easier to explain.

6.4 Bias-Variance Decomposition

Over-training (or overfitting) a particular algorithm is an expression used to describe the complexity of this algorithm, and hence its capacity, to fit the current training (e.g. getting a very small value of the RSS). Although this seems a success, it is not! This is because what is required is to have the best performance on the unseen data (the population of testers, as opposed to the training dataset itself). As explained in Sect. 5.2, when a ML algorithm trains on a training dataset there are two things to realize: (1) the training dataset is taken as an example of the population, and (2) the objective function, to be minimized on this training dataset, is an estimator of the performance measure that we hope to be minimized on the population. For example, recall that we used the RSS as an objective function to estimate the MSE as a performance measure.

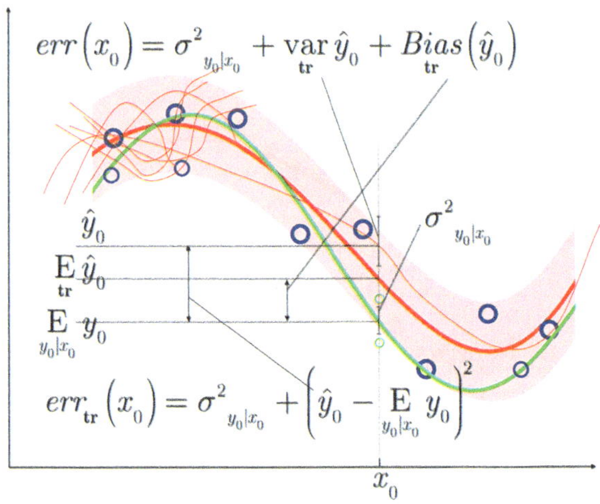

Fig. 13 Bias-variance decomposition. A visual illustration using the same colors of the mathematical quantities in Eq. (51). (Part of the background of the figure, namely, the bold green curve, the bold red curve, and the shaded area, is as appears in [6, Fig. 1.17, p. 32])

Overfitting an algorithm results in decreasing the bias of the performance measure and increasing its variance, and vice versa; and there is always a trade-off between this bias and variance. Before delving into any mathematics, Fig. 12 qualitatively illustrates this phenomenon for two different ML algorithms (left and right). The bold function η^*, in both subfigures, is the conditional expectation, which is the best regression function. Training the algorithm on a training dataset **tr** produces a function $\eta_{\mathbf{tr}}$ that may exist anywhere in the shaded region in the figure. The pointwise mean of these functions is plotted in light black. The algorithm of the left subfigure produces a mean model $\mathrm{E}_{\mathbf{tr}}\, \eta_{\mathbf{tr}}(x)$ that is very close to η^* (low bias); however, a single fitted model may exist anywhere in the wide shaded region (high variance). The algorithm of the right subfigure behaves conversely.

This can be best understood if the KNN is taken as an example. At some point x_i, the prediction is $\Sigma_{j \in N_K(x_i)} y_j / K$. The expectation of this regression function is $\Sigma_{j \in N_K(x_i)} \mathrm{E}\left[y_j\right] / K$, while the variance will be σ^2/K (where the response is assumed to have constant variance σ^2 with the predictor). If the window size of this rule is squeezed to produce a more complex rule, i.e., K is decreased, the variance will increase, but the bias will decrease since $\Sigma_{j \in N_K(x_i)} \mathrm{E}\left[y_j\right] / K$ tends to approach $\mathrm{E}\left[y_i\right]$. On the contrary, increasing K obviously decreases the variance, while incorporating many data points whose expectations will be very likely to vary from $\mathrm{E}\left[y_i\right]$, hence the bias increases. This example of KNN is provided in [20]. For more elaboration [19, Chap. 3], review a measure of the complexity of smoothing functions in terms of an effective number of degrees of freedom.

The bias-variance decomposition for a regression function is analyzed quantitatively in Eq. (51), and is illustrated in Fig. 13. This figure is conceptually similar to Fig. 12, but with indicating each component of Eq. (51) on the figure. For better illustration and pedagogy, the colors of the figure match the colors of the corresponding terms of the equation as follows: the light red for a trained model, the bold red for its mean over the population of training datasets, the blue circles for the train-

$$\overline{\text{Err}} = \frac{1}{n_{\mathbf{tr}}} \sum_{i \in \mathbf{tr}} (\widehat{y}_i - y_i)^2, \tag{51a}$$

$$\text{Err}_{\mathbf{tr}} = \text{E}_{x_0, y_0} \left[(\widehat{y}_0 - y_0)^2 \right], \tag{51b}$$

$$= \text{E}_{x_0} \left[\text{E}_{y_0|x_0} (\widehat{y}_0 - y_0)^2 \right] \qquad (\cong \frac{1}{n_{\mathbf{ts}}} \sum_{i \in \mathbf{ts}} (\widehat{y}_i - y_i)^2), \tag{51c}$$

$$= \text{E}_{x_0} \left[\sigma^2_{Y|X=x_0} + \left(\widehat{y}_0 - \underset{y_0|x_0}{\text{E}} y_0 \right)^2 \right], \tag{51d}$$

$$\text{Err} = \text{E}_{\mathbf{tr}} \, \text{Err}_{\mathbf{tr}} \qquad (\cong \frac{1}{M} \sum_{m=1}^{M} \text{Err}_{\mathbf{tr}_m}), \tag{51e}$$

$$= \text{E}_{x_0} \left[\underbrace{\sigma^2_{Y|X=x_0}}_{\text{response variance}} + \underbrace{\text{E}_{\mathbf{tr}} \left(\widehat{y}_0 - \underset{\mathbf{tr}}{\mathbf{E}} \widehat{y}_0 \right)^2}_{\text{Variance}} + \underbrace{\left(\underset{\mathbf{tr}}{\mathbf{E}} \widehat{y}_0 - \underset{y_0|x_0}{\text{E}} y_0 \right)^2}_{\text{Bias squared}} \right]. \tag{51f}$$

ing dataset (observed response), the green circles for the testing dataset (observed response), the bold green for the response conditional mean (the best regression function). The symbol $\cong$ is used to indicate how the corresponding quantity can be estimated from a dataset, regardless whether this is a good estimator or not as will be explained in the next chapter. The symbol M denotes the number of training datasets drawn through Monte Carlo (MC) trials.

We end this discussion with the following two questions that pave the road for the subfield of *ML assessment*, the topic of the next chapter. These questions are valid for any other performance measure, and the error rate is given just as a clear example. The subfield of *ML assessment* explains different methods for estimating the performance of a ML algorithm, from a given dataset (because the whole population of testers is unknown) to select among a variety of competing models and assess them, which answers the following two questions.

(1) How can we minimize the mean error $\text{E}_{\mathbf{tr}} \, \text{Err}_{\mathbf{tr}}$ (51f), where the expectation is taken over the population of training datasets? As appears from the equation, this error is decomposed to three terms: the response variance, the model variance, and the model squared bias, respectively. The response variance is the natural variance in the physical phenomenon that generated the data and is model independent; hence, it is irreducible. Therefore, to minimize the mean error $\text{E}_{\mathbf{tr}} \, \text{Err}_{\mathbf{tr}}$, the model complexity should be tuned so that the summation of the bias squared and variance is minimized. Tersely speaking: too simple (complex) models produce high bias (variance) and low variance (bias); and since the variance (bias squared) cannot drop below zero, the summation of these two quantities will be high.

(2) Should we design the ML model to minimize the conditional error $\text{Err}_{\mathbf{tr}}$ (conditional on a particular training dataset) or the mean error $\text{E}_{\mathbf{tr}} \, \text{Err}_{\mathbf{tr}}$ that involves the bias and variance components?

6.5 Curse of Dimensionality

This expression refers to what may happen when the predictor has high dimensions, i.e., p is too large. The word "large" should be understood relatively to the size n of the available dataset. For illustration, consider smoothing in high dimensions. It will almost fail because for a fixed number of available observations (the training dataset), the volume size needed to cover a particular percentage of the total number of observations increases by a power law, and thus exponentially, with dimensionality. This makes it prohibitive to include the same sufficient number of observations within a small neighborhood, or bandwidth, to smooth the response. More quantitatively, consider a unit hyper-cube in the p-dimensional space containing uniformly distributed observations; the percentage of the points located inside a hyper-cube with side length l is l^p. This means, if the suitable bandwidth for a certain smoother is l, the effective number of observations in the p-dimensional problem will go as the power $1/p$. This deteriorates the performance dramatically for $3 < p$. This is why, e.g., the additive model (Sect. 4.2) and its variants are expressed as summation of functions of just one dimension. This single dimension may be just a component of the predictor or a linear combination.

Many other problems occur when $n \ll p$, including increasing the model variance. In addition, the performance of the model fitted from a given dataset will not generalize on the population or a future dataset. All of these problems, and others, are indeed related and connected mathematically to each other, which is out of the scope of the present chapter.

Therefore, a very crucial topic in ML is dimensionality reduction (it is called feature selection in some other communities). Qualitatively speaking, this means selecting those predictor components that best summarize the relationship between the response and predictor. In real-life problems, some features are statistically dependent on others; this is referred to as multi-collinearity. On the other hand, there may also be some components that are statistically independent from the response. These add no additional information to the problem; thus they serve only as a source of noise.

Several existing approaches aim to reduce the dimensions of the problem. A dimensionality reduction method of course can be considered as part of the ML algorithm. Therefore, for a given problem, selecting among different methods account as selecting among different algorithms which is the main topic of the next chapter, as explained at the end of Sect. 6.4.

6.6 Performance of Unsupervised Learning

It should be noticed that the formal definition of the learning process, discussed thus far in the present chapter, assumed the existence of a training dataset, $\mathbf{tr} : \{t_i = (x_i, y_i)\ ,\ i = 1, \ldots, n\}$. Each element t_i, or sample case, in this set has an already known value for the response variable. This is what enables the learning process to develop the relationship between the predictor and the response. This is what is called supervised learning. On the contrary, in some applications the available dataset is described by $\mathbf{tr} : \{t_i = x_i,\ i = 1, \ldots, n\}$, without any additional information. This situation is called unsupervised learning. It is usually required in such a situation to understand the structure of the data from the available empirical probability distribution of the points x_i. For the special case, where the data come from different classes, the data will be represented in the hyper p-dimensional space, to some extent, as disjoint clouds of data. The task in this case is called clustering, i.e., trying to identify those classes that best describe, in some sense, the current available data. More formally, if the available dataset is $\mathbf{X}$, it is required to find the class vector $\Omega = [\omega_1, \ldots, \omega_k]'$ and the clustering function $\eta_{\mathbf{tr}(X)}$, such that a criterion (an objective function) $\mathscr{J}(\mathbf{X}, \Omega)$ is minimized:

$$\Omega = \arg\min\left[\mathscr{J}(\mathbf{X}, \Omega)\right]. \tag{51}$$

Different criteria give rise to different clustering algorithms. More discussion on unsupervised learning and clustering can be found in [13, 17, 20].

It is important to emphasize that although the construction of the supervised and unsupervised rules is quite distinct, the assessment procedure and the performance measures, including error rate, risk, ROC, AUC, etc., are essentially the same. This is obvious because the unsupervised rule $\Omega_{\mathbf{tr}}(X)$, regardless of its construction, ultimately provides the same mapping $\eta_{\mathbf{tr}}(X) \mapsto \{\omega_1, \ldots, \omega_K\}$ as the supervised rule, which is assigning a class label to a predictor.

6.7 *Classifier Calibration*

As detailed throughout the chapter, the final classifier decision $\eta_{\mathbf{tr}}(x)$ of a classifier is obtained by comparing its output score $h_{\mathbf{tr}}(x)$ to a threshold th. However, there are two important issues to consider. (1) The scores do not necessarily equal to the posterior probabilities $\Pr[\omega_i|x]$, which are much more informative than a mere numerical score; indeed, many classifiers provide score values outside the period [0, 1]. (2) Scores of two different classifiers cannot be compared, simply because they are not on the same scale. Classifier calibration is a remedy to these two issues, not naively by linear scaling, but by providing a one-to-one nonlinear monotonic transformation that maps the output scores to the posterior probabilities. It is important to observe that this transformation will not affect the performance of the classifier on the population or on a finite testing dataset. For a formal proof of this result, and for a full account of the calibration process including a recent comparative study among different calibrators see [29].

7 Discussion and Conclusion

This chapter is intended to provide a pilot view of the field of ML to illustrate how mathematics and intuition together work, which helps cyberphysical security practitioners, who apply ML in many applications, understand subtle concepts and connect scattered pieces. The importance of the theoretical aspects of ML are stressed, and demonstrating examples are provided. The mathematical foundations of the field, along with different methods and construction, have been motivated. Important and fundamental references have been cited for readers, who are interested in more elaboration.

When it comes to real-life applications, many practitioners leverage some ML approaches, or models, without having the fundamental rigour or the enough insight, a matter that results in a lot of fallacies and pitfalls. A simple example is the use of complex models, that have high capacity, relative to the training dataset size. A second example is to perform data preprocessing or transformation without including the step into the resampling mechanism that estimates the final performance. A third example is thinking of a particular model or approach as "magical" that can consistently outperform others ubiquitously.

"No overall winner" is a statement that has been touched upon throughout previous sections. If there is no prior information for the joint distribution between the response and the predictor, and if there is no prior information about the phenomenon to which that regression or classification will be applied, there is no overall winner among regression or classification techniques. If one method is found to outperform others in some applications, this is likely to be limited to that very situation or that specific kind of problem; it may be beaten by other methods for other situations. In the engineering and computer science communities, this concept is referred to as the *no-free-lunch* theorem (see [13, Sect. 9.2]). This situation holds because each method makes different assumptions about the application or the process being modeled, and not all real-life applications are the same. If one or more of the assumptions are not satisfied in a given application, the performance will not be optimal in that setting. The only unique overall winner is the conditional expectation (for regression) or the Bayes' classifier (for classification) when the probability distributions are known.

To recap, practitioners are always advised to have a basic level of mathematical rigor and understanding of these foundations, even if they do not produce research or contribute to theoretical discoveries in the field.

Acknowledgements The author is grateful to the U.S. Food and Drug Administration (FDA) for funding a very early stage of this chapter, and to Dr. Kyle Myers for her support. In his memorial, special thanks and gratitude to Dr. Robert F. Wagner, the supervisor and the teacher, or Bob Wagner, the big brother and friend. He reviewed a very early version of this chapter before he passed away.

References

1. Anderson TW (2003) An introduction to multivariate statistical analysis, 3rd edn. Wiley, Hoboken, NJ
2. Bezdek JC (1992) On the relationship between neural networks, pattern recognition and intelligence. Int J Approx Reason 6:85–107
3. Bezdek JC (1994) What is computational intelligence? In: Zurada JM, Marks RJ, Robinson CJ (eds) Computational intelligence: imitating life. New York, pp 1–12
4. Billingsley P (1995) Probability and measure, 3rd edn. Wiley, New York
5. Bishop CM (1995) Neural networks for pattern recognition. Clarendon Press, Oxford University Press, Oxford, New York
6. Bishop CM (2006) Pattern recognition and machine learning. Springer, New York
7. Bowerman BL, O'Connell RT (1990) Linear statistical models: an applied approach, 2nd edn. PWS-Kent Pub., Co., Boston
8. Boyd S, Vandenberghe L (2004) Convex optimization. Cambridge University Press, Cambridge, https://doi.org/10.1017/CBO9780511804441
9. Casella G, Berger RL (2002) Statistical inference, 2nd edn. Duxbury/Thomson Learning, Australia; Pacific Grove, CA
10. Cherkassky VS, Mulier F (1998) Learning from data: concepts, theory, and methods. Wiley, New York
11. Chong EK, Zak H Stanislaw (2013) An introduction to optimization, 4th edn. Wiley
12. Christensen R (2002) Plane answers to complex questions: the theory of linear models, 3rd edn. Springer, New York
13. Duda RO, Hart PE, Stork DG (2001) Pattern classification, 2nd edn. Wiley, New York
14. Efron B (1975) The efficiency of logistic regression compared to normal discriminant analysis. J Am Stat Assoc 70(352):892–898
15. Engelbrecht AP (2002) Computational intelligence: an introduction. Wiley, Chichester, England; Hoboken, NJ
16. Friedman JH, Stuetzle W (1981) Projection pursuit regression. J Am Stat Assoc 76(376):817–823
17. Fukunaga K (1990) Introduction to statistical pattern recognition, 2nd edn. Academic Press, Boston
18. Graybill FA (1976) Theory and application of the linear model. Duxbury Press, North Scituate, Mass
19. Hastie T, Tibshirani R (1990) Generalized additive models, 1st edn. Chapman and Hall, London, New York
20. Hastie T, Tibshirani R, Friedman JH (2009) The elements of statistical learning: data mining, inference, and prediction, 2nd edn. Springer, New York
21. Nadaraya EA (1964) On estimating regression. Theory Prob Appl 9(1):141–142
22. Parzen E (1962) On estimation of a probability density function and mode. Ann Math Stat 33(3):1065–1076
23. Rencher AC (2000) Linear models in statistics. Wiley, New York
24. Ripley BD (1996) Pattern recognition and neural networks. Cambridge University Press, Cambridge, New York
25. Schwefel HP, Wegener I, Weinert K (2003) Advances in computational intelligence: theory and practice
26. Watson ES (1964) Smooth regression analysis. Sankhyā: Ind J Stat Ser A 359–372
27. Yousef WA (2013) Assessing classifiers in terms of the partial area under the roc curve. Comput Stat Data Anal 64:51–70
28. Yousef WA (2020) Prudence when assuming normality: an advice for machine learning practitioners. Pattern Recogn Lett 138:44–50
29. Yousef WA, Traore I, Briguglio W (2021a) Classifier calibration: with application to threat scores in cybersecurity. arXiv:2102.05143, https://github.com/isotlaboratory/ClassifierCalibration-Code

30. Yousef WA, Traoré I, Briguglio W (2021b) UN-AVOIDS: unsupervised and nonparametric approach for visualizing outliers and invariant detection scoring. IEEE Trans Inf Forensics Secur 16:5195–5210, https://doi.org/10.1109/TIFS.2021.3125608
31. Zimmermann HJ, Tselentis G, van Someren M, Dounias D (2002) Advances in computational intelligence and learning: methods and applications

Machine Learning Assessment: Implications to Cybersecurity

Waleed A. Yousef

Abstract After discussing the construction of machine learning (ML) algorithms in the previous chapter, this chapter is dedicated to their assessment and performance estimation (with an emphasis on classification assessment), a topic that is equally important specially in the context of cyberphysical security design. The literature is full of nonparametric methods to estimate a statistic from just one available dataset through resampling techniques, e.g., jackknife, bootstrap and cross validation (CV). Special statistics of great interest are the error rate and the area under the ROC curve (AUC) of a classification rule. The importance of these resampling methods stems from the fact that they require no knowledge about the probability distribution of the data or the construction details of the ML algorithm. This chapter provides a concise review of this literature to establish a coherent theoretical framework for these methods that can estimate both the error rate (a one-sample statistic) and the AUC (a two-sample statistic). The resampling methods are usually computationally expensive, because they rely on repeating the training and testing of a ML algorithm after each resampling iteration. Therefore, the practical applicability of some of these methods may be limited to the traditional ML algorithms rather than the very computationally demanding approaches of the recent deep neural networks (DNN). In the field of cyberphysical security, many applications generate structured (tabular) data, which can be fed to all traditional ML approaches. This is in contrast to the DNN approaches, which favor unstructured data, e.g., images, text, voice, etc.; hence, the relevance of this chapter to this field.

Keywords Assessment · Performance estimation · Resampling techniques · ROC curve · Area under the ROC curve · Classification · Machine learning · Deep neural network · Unstructured data · Sample · Bootstrap · Nonparametric · Estimators

---

W. A. Yousef (✉)
CS Department, HCILAB, Faculty of Computers and Artificial Intelligence, Helwan University, Helwan, Egypt
e-mail: wyousef@fci.helwan.edu.eg

I. Traore et al. (eds.), *Artificial Intelligence for Cyber-Physical Systems Hardening*, Engineering Cyber-Physical Systems and Critical Infrastructures 2,
https://doi.org/10.1007/978-3-031-16237-4_3

1 Introduction

1.1 Motivation

Consider a ML problem, where some models have been trained on a given dataset. It is then required to know their performances, in terms of any performance measure, on the population of testers. This is not only for the sake of assessing each of them, but also to be able to select the best model among them. These different models could even represent different instances of the same ML algorithm, with different values of parameters (e.g., a KNN with different values of K), and it is required to choose the best value for the current problem. The performance on the population of testers is called the true performance, because this is the performance on the whole population, not on a subset of it.

If the underlying probability distribution of the testers is known, e.g., from a priori knowledge about the nature of the problem, the true performance can be calculated mathematically. One of the first attempts in this direction was Fukunaga [14], where he assumed the data follows a multinormal distribution, to find a closed-form expression of the error rate of a binary classification rule. An alternative to mathematical calculations is simulating a very large dataset, from the assumed distribution, from which a very accurate estimation of the true performance can be obtained.

The early work of Fukunuga was inspiring, from the theoretical point of view, for the early community of pattern recognition and machine learning to understand important theoretical properties and concepts. However, for real-life applications it is very unusual that the assumption of multinormality, or any other assumption, hold. In these situations, which are called nonparametric, or distribution-free, it is impossible to derive either the true performance in closed form, or estimate it using a very large simulated dataset. In such situations, the true performance must be estimated from a single testing dataset (testers). The way we obtain such a testing dataset defines two major paradigms, discussed next.

In Paradigm I, we only have one dataset $\mathbf{t}$, usually called the design or construction dataset, from which we have to make up a training dataset $\mathbf{tr}$ and a testing dataset $\mathbf{ts}$, such that $\mathbf{t} = \mathbf{tr} \cup \mathbf{ts}$. Otherwise, training and testing on the same dataset $\mathbf{t}$ would provide a very optimistic estimate of the performance measure. This splitting is performed iteratively using one of the resampling techniques, e.g., jackknife, bootstrap, or cross validation. In each resampling iteration we get a different pair of training and testing datasets, on which the algorithm will be trained and tested, respectively. The results from these different iterations will be compiled together, as defined by the resampling method, to provide a single estimate of the performance measure. It is obvious that the performance estimation obtained from any of these methods will vary with varying the design dataset $\mathbf{t}$. This chapter is dedicated to reviewing this paradigm, its different estimators, and the variance estimation of these estimators.

It is worth mentioning that fatal fallacies are committed by practitioners when using this paradigm. For example, a very common mistake is using the whole dataset $\mathbf{t}$ to learn some statistical properties of the different classes of the classification problem, mistakenly naming this a data preprocessing step, using these properties to construct a classifier, then excluding this step from the resampling mechanism afterwards. Although the correct way of performing preprocessing is explained in textbooks (see, e.g.,Hastie et al. [20], Sect. 7.10.2), we still see this mistake in several occasions in both academia and industry.

In Paradigm II, it is required, or even mandated (e.g., in several public-policy-making or regulatory settings), to maintain what might be called the traditional data hygiene of two independent datasets: the design dataset **t**, and a final testing dataset **TS**, which is a sequestered testing dataset that has never been available to the design procedure, but for just onetime final testing. Assessing a ML algorithm from independent testing dataset is as simple as applying the estimators of the performance measure of interest (Sect. 1.2) on the testing dataset. However, the estimator will then have two sources of variability, the design and the testing datasets. The mathematical details of this paradigm and the estimation of this variance are discussed in Yousef et al. [34], Chen et al. [4], and not reviewed in our present chapter.

Although it may seem very safe to use this testing paradigm, some practitioners abuse it as well. One possible common mistake is that they test several models on this sequestered testing set, then they analyze the relative estimated performances. Accordingly, these models are redesigned to improve their performance on the testing set! Worse than this is keeping iterating this processes several times, which indeed turns the independent sequestered testing dataset to be part of the training dataset, indirectly through this human mental parsing of the results, which acts as a feedback that guides the redesign process.

Nowadays, it is almost the default in the field of ML to leverage both paradigms in the task of model assessment and selection. The available dataset is initially split into two datasets:

1. the design dataset **t**, from which the ML algorithm is designed. This is conducted via one of the resampling methods of paradigm I explained above. Usually, several algorithms are used, and several parameters' values are examined for each algorithm. Then, the model with the best performance is chosen.
2. the sequestered testing dataset **TS**, on which the final chosen model from paradigm I is assessed once and only once, without redesign. This is the final estimation of the performance measure that should be reported, along with the estimation of its variance.

It is worth mentioning that, there is a convention in the field to call the dataset **ts** that is split from the design dataset **t** during the resampling process, a validation dataset rather than a testing dataset, to reserve the word testing to the final testing datset **TS** of paradigm II. However, in some applications, the converse is adopted; i.e., **ts** is called the a testing dataset and **TS** is called a validation dataset. To avoid ambiguity, any notation and expression should be defined clearly within any context.

What is introduced above is valid for any ML problem, whether it is regression or classification, and for any performance measure, whether it is the error rate Err, AUC, or any other. However, we emphasize below two very important issues.

(1) The true performance, which we discussed its estimation in this introduction so far, is itself a random variable whose randomness arises from the randomness of

the training dataset, as was explained in the previous chapter. Have we changed the training dataset, the true performance would change. For example, and without loss of generality (WLOG) but for the sake of illustration, suppose the whole design dataset **t** is used as a training dataset **tr** and we are interested in the AUC as a performance measure. Then, as was explained in the previous chapter, we should be interested in the following:

1. $\mathrm{AUC}_{\mathbf{t}}$: the true performance conditional on a particular training dataset **t** of a specified size n.
2. $\mathrm{E}_{\mathbf{t}}\mathrm{AUC}_{\mathbf{t}}$: the expectation of true performance over the population of training datasets of the same size n.
3. $\mathrm{Var}_{\mathbf{t}}\mathrm{AUC}_{\mathbf{t}}$: the variance of the true performance over the population of training datasets of the same size n.

(2) Regarding the meaning and utility of the performance measure, we emphasize the importance of the ROC curve and its AUC as a summary measure [2, 18, 19], where the former is a manifestation of the trade-off between the two types of error of any binary classification rule. We always advocate for the use of the ROC or its AUC since they are prevalence independent; i.e., they do not depend on a particular chosen threshold, class prior probability, or misclassification costs. Adopting a performance measure that is prevalence dependent, e.g., the overall accuracy or its many different versions, can provide a misleading measure of the classification power of the classification algorithm, especially in classification problems that involve, for instance, unbalanced data (different class size). Therefore, the present chapter assumes familiarity with the ROC and its AUC, at the level provided in the previous chapter. However, for the sake of completeness, all notations are tersely summarized in the following subsection.

1.2 Notation

Consider the binary classification problem, where a classification rule η gives a score of $h(x)$ for the predictor x, and classifies it to one of the two classes by comparing this score $h(x)$ to a chosen threshold th. The observation x belongs to one of the two classes with distributions $F_i, i = 1, 2$. The two error components of this rule (e_1, or the false negative fraction (FNF), and e_2 or the false positive fraction (FPF)), along with the risk, are given as follows:

$$\mathrm{FNF} = e_1 = \int_{-\infty}^{th} f_h\left(h(x)|\omega_1\right) dh(x), \tag{1a}$$

$$\mathrm{FPF} = e_2 = \int_{th}^{\infty} f_h\left(h(x)|\omega_2\right) dh(x), \tag{1b}$$

$$\mathrm{R} = c_{12} P_1 e_1 + c_{21} P_2 e_2. \tag{1c}$$

The cost $c_{ij},\ i, j = 1, 2$ is the cost of classifying an observation as belonging to class j whereas it belongs to class i; $c_{ii} = 0$, which means there is no cost for correct classification; and P_i is the prior probability of each class, $i = 1, 2$. The risk (1c) is called the "error rate" Err, or probability of misclassification (PMC), when putting $c_{12} = c_{21} = 1$, which is denoted by the 0-1 cost, or loss.

The receiver operating characteristics (ROC) curve is a plot of the true positive fraction (TPF), which is $1 - \mathrm{FNF}$, versus the FPF. Then the area under the curve (AUC) is given by:

$$\mathrm{AUC} = \int_0^1 \mathrm{TPF}\ d(\mathrm{FPF}). \tag{2a}$$

$$= \Pr\left[h(x)|\omega_2 < h(x)|\omega_1\right], \tag{2b}$$

which expresses how the classifier scores for class ω_1 are stochastically larger than those of class ω_2.

If the distributions F_1 and F_2 are not known, a setup that is called nonparametric or distribution-free, any performance measure can be estimated only numerically from a given dataset, called the testing dataset. This is regardless of the testing paradigm, i.e., whether this testing dataset is obtained by simulation, resampling, or sequestering. This is done by assigning equal probability mass for each observation:

$$\hat{F} : \text{mass } \frac{1}{n} \text{ on } t_i,\ i = 1, \ldots, n, \tag{3}$$

where n is the size of the testing dataset. Lemma 1 shows that this is the maximum likelihood estimator (MLE) of the distribution F.

In this case the performance measures (1) can be obtained as follows.

$$\widehat{\mathrm{FNF}} = \widehat{e_1} = \frac{1}{n} \sum_{i=1}^{n} I_{h(x_i|\omega_1)<th} \tag{4a}$$

$$\widehat{\mathrm{FPF}} = \widehat{e_2} = \frac{1}{n} \sum_{i=1}^{n} I_{h(x_i|\omega_2)>th} \tag{4b}$$

$$\widehat{\mathrm{R}(\eta)} = \frac{1}{n} \left(c_{12}\, \widehat{e_1}\, n_1 + c_{21}\, \widehat{e_2}\, n_2\right). \tag{4c}$$

The indicator function I_{cond} equals 1 or 0 when the Boolean expression *cond* is true or false, respectively. The values n_1 and n_2 are the number of observations in the two classes respectively, and $\widehat{P_1}$ and $\widehat{P_2}$ are the estimated a priori probabilities for each class.

As the the two components TPF and FPF defined a single operating point on the ROC, the two components $\widehat{\text{TPF}}(= 1 - \widehat{\text{FNF}})$ and $\widehat{\text{FPF}}$ give one point on the empirical (estimated) ROC curve. To draw the complete curve in the nonparametric situation, the classifier's sore is calculated for each point of the available dataset. Then all possible thresholds are considered in turn, i.e., the threshold values between every two successive scores. At each threshold value a point on the ROC curve is calculated. Then the AUC (2a) can be estimated from the empirical ROC curve using the trapezoidal rule:

$$\widehat{\text{AUC}} = \frac{1}{2}\sum_{i=2}^{n_{th}} (\text{FNF}_i - \text{FNF}_{i-1})(\text{TPF}_i + \text{TPF}_{i-1}), \tag{5}$$

where n_{th} is the number of threshold values taken over the dataset. By plotting the empirical ROC curve, it is easy to see that (5) is the same as the Mann-Whitney statistic—which is another form of the Wilcoxon rank-sum test [15, Chap. 4]—defined by:

$$\widehat{\text{AUC}} = \frac{1}{n_1 n_2}\sum_{j=1}^{n_2}\sum_{i=1}^{n_1} \psi\left(h\left(x_i|\omega_1\right), h\left(x_j|\omega_2\right)\right), \tag{6a}$$

$$\psi(a, b) = \begin{cases} 1 & a > b \\ 1/2 & a = b \\ 0 & a < b \end{cases}. \tag{6b}$$

It is interesting, as well, to know from the theory of U-statistics [25] that the estimator (6) is the uniform minimum variance unbiased estimator (UMVUE) for the probability (2b) under the distribution (3).

All the estimators given above have the nice property of converging to their corresponding population definitions, (1) and (2), as the size of the testing set goes to infinity. It is worth mentioning that each of the error estimators $\hat{e}_1$ and $\hat{e}_2$ in (4) is called a one-sample statistic, because its kernel $I_{(\cdot)}$ requires only one observation from either distributions. However, the AUC estimator in (6) is a two-sample statistic since its kernel $\psi(\cdot, \cdot)$ requires two observations, one from each distribution. This is a fundamental difference between both estimators (statistics) which will be treated and explained carefully in the present chapter.

1.3 Roadmap

The rest of this chapter is organized as follows. Section 2 paves the road to the chapter by reviewing the nonparametric estimators for estimating the mean and variance of one-sample statistics, including the preliminaries of bootstraps and influence function. This section is a very concise review mainly of the work done in Hampel [16],

Efron and Tibshirani [11], and Huber [21]. Section 3 switches gears and reviews the nonparametric estimators that estimate the mean and variance of a special kind of statistics, i.e., the error rate of classification rules. This section is a concise review of the work done mainly in Efron [8], and Efron and Tibshirani [13]. Section 4 explains how the nonparametric estimators that estimate the error rate, a one-sample statistic, can be extended to estimate the AUC, a two-sample statistic. It does so by providing theoretical parallelism between the two sets of estimators and showing that the extension is rigorous and not just an ad hoc application. Section 6 concludes the chapter and provides a discussion and an advice for practitioners.

2 Nonparametric Methods for Estimating the Bias and the Variance of a Statistic

Consider a statistic s that is a function of a dataset $\mathbf{x} : \{x_i,\ i = 1, \ldots, n\}$, where $x_i \overset{i.i.d}{\sim} F$. The statistic s is now a random variable and its variability comes from the variability of x_i. Suppose that this statistic is used to estimate a real-valued parameter $\theta = f(F)$. Then $\hat{\theta} = s(\mathbf{x})$ has expected value E $s(\mathbf{x})$ and variance Var $s(\mathbf{x})$. The mean squared error (MSE) of the estimator $\hat{\theta}$ is defined as:

$$\mathrm{MSE}(\hat{\theta}) = \mathrm{E}\left[\hat{\theta} - \theta\right]^2. \tag{7}$$

The root of the mean squared error (RMS) has the same units and is on the same scale of the original variable θ, and hence has more intuitive value. The bias of the estimator $\hat{\theta} = s(\mathbf{x})$ is defined by the difference between the true value of the parameter and the expectation of the estimator, i.e.,

$$\mathrm{bias}_F\left(\hat{\theta}\right) = \mathrm{E}_F s(\mathbf{x}) - \theta. \tag{8}$$

Then, it is known that, the MSE in (7) can be decomposed to:

$$\mathrm{MSE}(\hat{\theta}) = \mathrm{bias}_F^2\left(\hat{\theta}\right) + \mathrm{Var}_F \hat{\theta}. \tag{9}$$

A critical question is whether the bias and variance of the statistic s in (9) may be estimated from the available dataset?

2.1 *Bootstrap Estimate*

The bootstrap was introduced by Efron [5] to estimate the standard error of a statistic. The bootstrap mechanism is implemented by treating the current dataset $\mathbf{x}$ as a

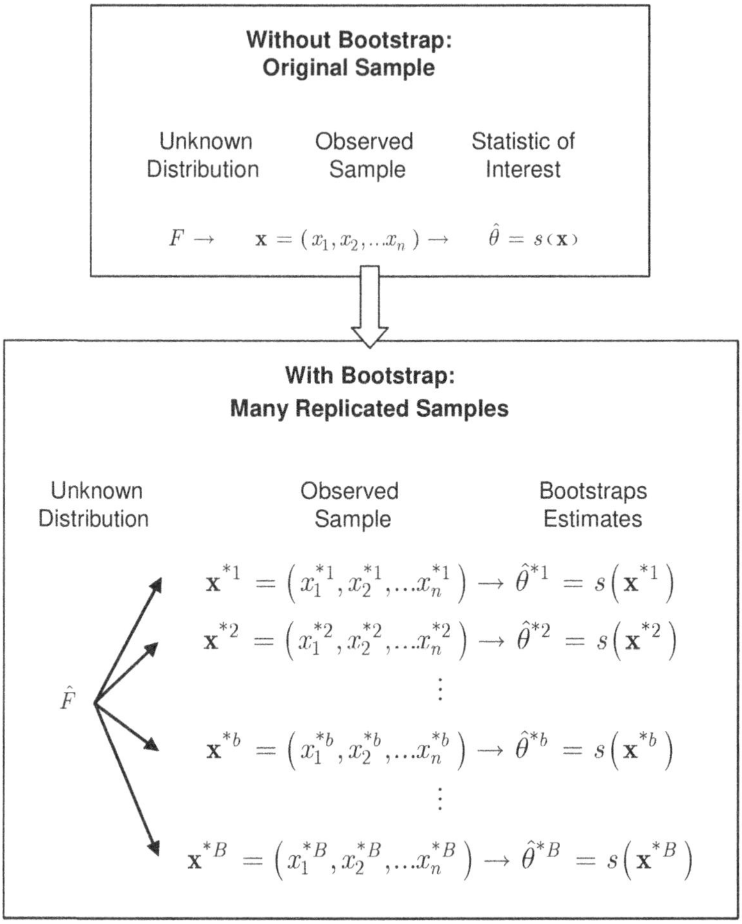

Fig. 1 Bootstrap mechanism: B bootstrap replicates are withdrawn (by sampling and replacement) from the original sample. From each replicate the statistic is calculated. (The idea behind this figure first appeared in [11, Fig. 6.1, pp. 48])

representation for the population distribution F; i.e., approximating the distribution F by the MLE defined in (3). Then B bootstrap samples are drawn from that empirical distribution. Each bootstrap replicate is of size n, the same size as $\mathbf{x}$, and is obtained by sampling with replacement. Then in a bootstrap replicate some case x_i, in general, will appear more than once at the expense of another x_j that will not appear. The original dataset will be treated now as the population, and the replicates will be treated as samples from the population. This situation is illustrated in Fig. 1. Each of these bootstrap replicates is denoted by $\mathbf{x}^{*b}$, $b = 1, \ldots, B$, and the corresponding bootstrap replications of the statistics $\hat{\theta} = s(\mathbf{x})$ itself are given by:

$$\hat{\theta}^{*b} = s(\mathbf{x}^{*b}), \quad b = 1, \ldots, B, \tag{10}$$

The bootstrap estimate of bias and standard error are defined by:

$$\text{bias}_B(\hat{\theta}) = \hat{\theta}^* - \hat{\theta}, \tag{11}$$

$$\widehat{\text{SE}}_B = \left[\frac{1}{(B-1)}\sum_{b=1}^{B}\left[\hat{\theta}^{*b} - \hat{\theta}^*\right]^2\right]^{1/2}, \tag{12}$$

$$\hat{\theta}^* = \frac{1}{B}\sum_{b=1}^{B}\hat{\theta}^{*b}. \tag{13}$$

Either in estimating the bias or the standard error, the larger the number of bootstraps B the closer the estimate to the asymptotic value, i.e.,

$$\lim_{B\to\infty} \widehat{\text{SE}}_B(\hat{\theta}^*) = \text{SE}_{\hat{F}}(\hat{\theta}^*). \tag{14}$$

For more details and some examples the reader is referred to [11, Chap. 6, 7, and 10].

2.2 *Jackknife Estimate*

Instead of replicating from the original dataset, a new set $\mathbf{x}_{(i)}$ is created by removing the case x_i from the dataset. Then the jackknife samples are defined by:

$$\mathbf{x}_{(i)} = (x_1, \ldots, x_{i-1}, x_{i+1}, \ldots, x_n), \quad i = 1, \ldots, n, \tag{15}$$

and the n-jackknife replications of the statistic $\hat{\theta}$ are:

$$\hat{\theta}_{(i)} = s(\mathbf{x}_{(i)}), \quad i = 1, \ldots, n. \tag{16}$$

The jackknife estimates of bias and standard error are defined by:

$$\widehat{\text{bias}}_J = (n-1)(\hat{\theta}^J - \hat{\theta}), \tag{17}$$

$$\widehat{\text{SE}}_J = \left[\frac{n-1}{n}\sum_{i=1}^{n}(\hat{\theta}_{(i)} - \hat{\theta}^J)^2\right]^{1/2}, \tag{18}$$

$$\hat{\theta}^J = \frac{1}{n}\sum_{i=1}^{n}\hat{\theta}_{(i)}. \tag{19}$$

For motivation behind the factors $(n-1)$ and $(n-1)/n$ in (17) see [11, Chap. 11]. The jackknife estimate of variance is discussed in detail in Efron [6] and Efron and Stein [10].

2.3 Bootstrap Versus Jackknife

Usually, it requires up to 200 bootstraps to yield acceptable bootstrap estimates; (in special situations like estimating the uncertainty in classifier performance it may take up to thousands of bootstraps). Hence, this requires calculating the statistic $\hat{\theta}$ the same number of times B, as well. In the case of the jackknife, it requires only n calculations as shown in (16). If the sample size is smaller than the required number of bootstraps, the jackknife is more economical in terms of computational cost.

In terms of accuracy, the jackknife can be seen to be an approximation to the bootstrap when estimating the standard error of a statistic [11, Chap. 20]. Thus, if the statistic is linear they almost give the same result; (the bootstrap gives the jackknife estimate multiplied by $[(n-1)/n]^{1/2}$). A statistic $s(\mathbf{x})$ is said to be linear if:

$$s(\mathbf{x}) = \mu + \frac{1}{n}\sum_{i=1}^{n}\alpha(x_i), \tag{20}$$

where μ is a constant and $\alpha(\cdot)$ is a function. This also can be viewed as having one data point at a time in the argument of the function α. Similarly, the jackknife can be seen as an approximation to the bootstrap when estimating the bias. If the statistic is quadratic, they almost agree except in a normalizing factor . A statistic $s(\mathbf{x})$ is quadratic if:

$$s(\mathbf{x}) = \mu + \frac{1}{n}\sum_{1\le i\le n}\alpha(x_i) + \frac{1}{n^2}\sum_{1\le i<j\le n}\beta(x_i, x_j). \tag{21}$$

An in-depth treatment of the bootstrap and jackknife, and their relation to each other, in mathematical detail is provided by Efron [7, Chaps. 1–5].

If the statistic is not smooth the jackknife will fail. Informally speaking, a statistic is said to be smooth if a small change in the data leads to a small change in the statistic. An example of a non-smooth statistic is the median. If the sample cases are ranked and the median is calculated, it will not change when a sample case changes unless this sample case bypasses the median value. Using the same argument, we can see that an example of a smooth statistic is the sample mean.

2.4 Influence Function, Infinitesimal Jackknife, and Estimate of Variance

The infinitesimal jackknife was introduced by Jaeckel [22]. The concept of the influence curve was introduced later by Hampel [16]. In the present context and for pedagogical purposes, the influence curve will be explained before the infinitesimal jackknife, since the former can be understood as the basis for the latter.

Following Hampel [16], let $\mathfrak{R}$ be the real line and s be a real-valued functional defined on the distribution F, which is defined on $\mathfrak{R}$. The distribution F can be perturbed by adding some probability measure (mass) on a point x. This should be balanced by a decrement in F elsewhere, resulting in a new probability distribution $G_{\varepsilon,x}$ defined by:

$$G_{\varepsilon,x} = (1-\varepsilon)F + \varepsilon\delta_x, \; x \in \mathfrak{R}. \tag{22}$$

Then, the influence curve $IC_{s,F}(\cdot)$ is defined by:

$$IC_{s,F}(x) = \lim_{\varepsilon \to 0^+} \frac{s\left((1-\varepsilon)\,F + \varepsilon\delta_x\right) - s\,(F)}{\varepsilon}. \tag{23}$$

It should be noted that F does not have to be a discrete distribution. A simple example of applying the influence curve concept is to consider the expectation $s = \int x\, dF(x) = \mu$. Substituting back in (23) gives:

$$IC_{s,F}(x) = x - \mu. \tag{24}$$

The meaning of this formula is the following: the rate of change of the functional s with the probability measure at a point x is $x - \mu$. This is how the point x influences the functional s. The influence curve can be used to linearly approximate a functional s, along with its variance, which is similar to taking up to only the first-order term in a Taylor series expansion (Appendix 7.2).

It is important to state here that s should be a functional in $\hat{F}$ that is an approximation to F, as was initially assumed in (23). If for example the value of the statistic s changes if every sample case x_i is duplicated, i.e., repeated twice, this is not a functional statistic. An example of a functional statistic is the biased version of the variance estimate $\Sigma_i(x_i - \bar{x}_i)^2/n$, while the unbiased version $\Sigma_i(x_i - \bar{x}_i)^2/(n-1)$ is not a functional statistic. Generally, any approximation $s(\hat{F})$ to the functional $s(F)$, by approximating F by the MLE $\hat{F}$, obviously will be functional. In such a case the statistic $s(\hat{F})$ is called the plug-in estimate of the functional $s(F)$. Moreover, the influence function (IF) method for variance estimation is applicable only to those functional statistics whose derivative (73) exists. If that derivative exists, the statistic is called a smooth statistic; i.e., a small change in the dataset leads a small change in the statistic. For instance, although the median is a functional statistic in the sense that duplicating any sample case will result in the same value of the median, it is not smooth as described at the end of Sect. 2.3. A key reference for the IF is Hampel [17]. Appendix 7.2 shows an interesting connection to the jackknife estimate.

3 Nonparametric Methods for Estimating the Error Rate of a Classification Rule

The review provided in this section is a terse summary of the main work of Efron [8, 11, 13]. In the previous section the statistic, or generally speaking the functional, was a function of just one dataset. For a non-fixed design, i.e., when the predictors of the testing set do not have to be the same as the predictors of the training dataset, a slight clarification for the previous notations is needed. The classification rule trained on the training dataset **t** will be denoted as $\eta_{\mathbf{t}}$. Any new observation that does not belong to **t** will be denoted by $t_0 = (x_0, y_0)$. Therefore, the classification loss is given by $L(y_0, \eta_{\mathbf{t}}(x_0))$. Any performance measure conditional on that training dataset will be similarly subscripted. Thus, all the performance measures should be subscripted **t**; and hence the risk and the error rate (1) should be denoted by $\mathrm{R}_{\mathbf{t}}$ and $\mathrm{Err}_{\mathbf{t}}$, respectively. In the sequel, for simplicity and WLOG, the 0-1 loss function will be used. In such a case the conditional error rate will be given by:

$$\mathrm{Err}_{\mathbf{t}} = \mathrm{E}_{0F} L\left(y_0, \eta_{\mathbf{t}}\left(x_0\right)\right), \quad (x_0, y_0) \sim F. \tag{25}$$

The expectation E_{0F} is subscripted so to emphasize that it is taken over the observations $t_0 \notin \mathbf{t}$. If the performance is measured in terms of the error rate and we are interested in the mean performance, not the conditional one, then it is given by:

$$\mathrm{Err} = \mathrm{E}_{\mathbf{t}} \mathrm{Err}_{\mathbf{t}}. \tag{26}$$

The expectation $\mathrm{E}_{\mathbf{t}}$ is the expectation over the training dataset **t**, which would be the same if we had written E_F; for notation clarity the former is chosen.

Consider a classification rule $\eta_{\mathbf{t}}$ already trained on a training dataset **t**. A natural next question is, given that there is just a single dataset available, how to use this dataset in assessing the classifier performance as well? Said differently, how should one estimate, using only the available dataset, the true classification performance of a classification rule in predicting new observations; these observations are different from those on which the rule was trained. In this section, we will review the principal methods in the literature for estimating both the true error rate (25) and its mean (26) of a classification rule.

3.1 Apparent Error

The apparent error is the error of the fitted model when it is tested on the same training data. Of course it is downward biased with respect to the true error rate since it results from testing on the same information used in training [9]. The apparent error is defined by:

$$\overline{\mathrm{Err}}_{\mathbf{t}} = \mathrm{E}_{\hat{F}} L(y, \eta_{\mathbf{t}}(x)), \quad (x, y) \in \mathbf{t} \tag{27a}$$

$$= \frac{1}{n} \sum_{i=1}^{n} \left[I_{\hat{h}_{\mathbf{t}}(x_i|\omega_1)<th} + I_{\hat{h}_{\mathbf{t}}(x_i|\omega_2)>th} \right]. \tag{27b}$$

Overfitting a classifier to minimize the apparent error is not the goal. The goal is to minimize the true error rate (25) or its mean (26).

3.2 *Cross Validation (CV)*

The basic concept of CV, as a resampling approach, has been proposed in different articles since the mid-1930s. The concept simply leans on splitting the data into two parts; the first part is used in design (or training) without any involvement of the second part. Then the second part is used to test the designed procedure; this is to test how the designed procedure will behave for new datasets. Stone [28] is a key reference for CV that proposes different criteria for optimization.

CV can be used to assess the prediction error of a model or in model selection. The true error rate in (25) is the expected error rate for a classification rule if tested on the population, conditional on a particular training dataset **t**. This performance measure can be approximated by the leave-one-out CV (LOOCV) by:

$$\widehat{\mathrm{Err}}_{\mathbf{t}}^{cv1} = \frac{1}{n} \sum_{i=1}^{n} L\left(y_i, \eta_{\mathbf{t}^{(i)}}(x_i)\right), \quad (x_i, y_i) \in \mathbf{t}. \tag{28}$$

This is done by training the classification rule on the dataset $\mathbf{t}^{(i)}$ that does not include the case t_i; then testing the trained rule on that omitted case. This proceeds in "round-robin" fashion until all cases have contributed one at a time to the error rate. There is a hidden assumption in this mechanism: the training dataset **t** will not change very much by omitting a single case. Therefore, testing on the omitted observation one at a time accounts for testing approximately the same trained rule on n new cases, all different from each other and different from those the classifier has been trained on. Besides this LOOCV, there are other versions named K-fold (or leave-n/K-out). In such versions the whole dataset is split into K roughly equal-sized subsets, each of which contains approximately n/K observations. The classifier is trained on $K - 1$ subsets and tested on the left-out one; hence we have K iterations. It is clear that the LOOCV is a special case of the K-fold CV, where $K = n$.

It is of interest to assess this estimator to see whether it estimates the conditional true error $\mathrm{E}\left[\widehat{\mathrm{Err}}_{\mathbf{t}}^{cv1} - \mathrm{Err}_{\mathbf{t}}\right]^2$, with small MSE, as was designed or not. Many simulation results, e.g., Efron [8], show that there is only a very weak correlation between the CV estimator $\widehat{\mathrm{Err}}_{\mathbf{t}}^{cv1}$ and the conditional true error rate $\mathrm{Err}_{\mathbf{t}}$. This issue is discussed in mathematical detail in the excellent paper by Zhang [35]. Those other estimators that are based on resampling as well, and will be reviewed below, are shown to have this same attribute. This very interesting (and perhaps surprising) result means the following: whether the estimator is designed to estimate the conditional performance or the mean performance it indeed estimates the latter because of the weak correlation with the former.

3.3 Bootstrap Methods for Error Rate Estimation

The prediction error in (25) is a function of the training dataset $\mathbf{t}$ and the testing population F. Bootstrap estimation can be implemented here by treating the empirical distribution $\hat{F}$ as an approximation to the actual population distribution F. By replicating from that distribution one can simulate many training datasets $\mathbf{t}^{*b}$, $b = 1, \ldots, B$. For every replicated training dataset the classifier will be trained and then tested on the original dataset $\mathbf{t}$. This is the simple bootstrap (SB) estimator approach [11, Sect. 17.6] that was defined formally by:

$$\widehat{\mathrm{Err}}_{\mathbf{t}}^{SB} = \mathrm{E}_* \sum_{i=1}^{n} L(y_i, \eta_{\mathbf{t}^*}(x_i))/n, \quad \hat{F} \to \mathbf{t}^*. \tag{29}$$

It should be noted that this estimator no longer estimates the true error rate (25) because the expectation taken over the bootstraps mimics an expectation taken over the population of trainers, i.e., it is not conditional on a particular training dataset. Rather, the estimator (29) estimates the expected performance of the classifier $\mathrm{E}_F \mathrm{Err}_{\mathbf{t}}$. For a finite number of bootstraps, the expectation (29) can be approximated by:

$$\widehat{\mathrm{Err}}_{\mathbf{t}}^{SB} = \frac{1}{B} \sum_{b=1}^{B} \sum_{i=1}^{n} L\left(y_i, \eta_{\mathbf{t}^{*b}}(x_i)\right)/n. \tag{30}$$

3.3.1 Leave-One-Out Bootstrap (LOOB)

The previous estimator is obviously biased since the original dataset $\mathbf{t}$ used for testing includes part of the training data in every bootstrap replicate. Efron [8] proposed that, after training the classifier on every bootstrap replicate, it is tested on those cases in the set $\mathbf{t}$ that are not included in the training; this concept can be developed as follows. Equation (30) can be rewritten by interchanging the order of the double summation to give:

$$\widehat{\mathrm{Err}}_{\mathbf{t}}^{SB} = \frac{1}{n} \sum_{i=1}^{n} \sum_{b=1}^{B} L\left(y_i, \eta_{\mathbf{t}^{*b}}(x_i)\right) \Big/ B. \tag{31}$$

This equation is formally identical to (30) but it expresses a different mechanism for evaluating the same quantity. It says that, for a given point, the average performance over the bootstrap replicates is calculated; then this performance is averaged over all the n cases. Now, if every case t_i is tested only from those bootstraps that did not include it in the training, a slight modification of the previous expression yields the leave-one-out bootstrap (LOOB) estimator:

$$\widehat{\mathrm{Err}}_{\mathbf{t}}^{(1)} = \frac{1}{n}\sum_{i=1}^{n}\left[\sum_{b=1}^{B} I_i^b L\left(y_i, \eta_{\mathbf{t}^{*b}}(x_i)\right) \Big/ \sum_{b'=1}^{B} I_i^{b'}\right], \tag{32}$$

where the indicator function I_i^b equals one when the case t_i is not included in the training replicate b, and zero otherwise. Efron and Tibshirani [13] emphasized a critical point about the difference between this bootstrap estimator and the LOOCV. The CV tests on a given sample case t_i, having been trained just once on the remaining dataset. By contrast, the LOOB tests on a given sample case t_i using a large number of classifiers that result from a large number of bootstrap replicates that do not contain that sample. This results in a smoothed cross-validation-like estimator. We explained and elaborated on this smoothness property in Yousef [30].

3.3.2 The Refined Bootstrap (RB)

The SB and the LOOB, from their definitions, look like designed to estimate the mean true error rate (26) of a classifier. For estimating the true conditional error rate of a classifier, conditional on a particular training dataset, Efron [8] proposed to correct for the downward biased estimator $\overline{\mathrm{Err}}_{\mathbf{t}}$. Since the true error rate $\mathrm{Err}_{\mathbf{t}}$ can be written as $\overline{\mathrm{Err}}_{\mathbf{t}} + (\mathrm{Err}_{\mathbf{t}} - \overline{\mathrm{Err}}_{\mathbf{t}})$, then it can be approximated by $\overline{\mathrm{Err}}_{\mathbf{t}} + \mathrm{E}_F(\mathrm{Err}_{\mathbf{t}} - \overline{\mathrm{Err}}_{\mathbf{t}})$. The term $(\mathrm{Err}_{\mathbf{t}} - \overline{\mathrm{Err}}_{\mathbf{t}})$ is called the optimism. The expectation of the optimism can be approximated over the bootstrap population. Finally the refined bootstrap approach, as named in Efron and Tibshirani [11, Sect. 17.6], gives the estimator:

$$\widehat{\mathrm{Err}}_{\mathbf{t}}^{RB} = \overline{\mathrm{Err}}_{\mathbf{t}} + \mathrm{E}_*(\mathrm{Err}_{\mathbf{t}*}(\hat{F}) - \overline{\mathrm{Err}}_{\mathbf{t}*}), \tag{33}$$

where $\mathrm{Err}_{\mathbf{t}*}(\hat{F})$ represents the error rate obtained from training the classifier on all bootstrap replicates $\mathbf{t}^*$ and testing on the empirical distribution $\hat{F}$. This can be approximated for a limited number of bootstraps by:

$$\widehat{\mathrm{Err}}_{\mathbf{t}}^{RB} = \overline{\mathrm{Err}}_{\mathbf{t}} + \frac{1}{B}\sum_{b=1}^{B}\left[\sum_{i=1}^{n} L\left(y_i, \eta_{\mathbf{t}^{*b}}(x_i)\right)/n - \sum_{i=1}^{n} L\left(y_{ib}^*, \eta_{\mathbf{t}^{*b}}(x_{ib}^*)\right)/n\right]. \tag{34}$$

3.3.3 The 0.632 Bootstrap

If the concept used in developing the LOOB estimator, i.e., testing on cases not included in training, is used again in estimating the optimism described above, this gives the 0.632 bootstrap estimator. Since the probability of including a case t_i in the bootstrap $\mathbf{t}^{*b}$ is given by:

$$\Pr(t_i \in \mathbf{t}^{*b}) = 1 - (1 - 1/n)^n \approx 1 - e^{-1} = 0.632, \tag{35}$$

the effective number of sample cases contributing to a bootstrap replicate is approximately 0.632 of the size of the training dataset. Efron [8] introduced the concept of a *distance* between a point and a sample in terms of a probability. Having trained on a bootstrap replicate, testing on those cases in the original dataset not included in the bootstrap replicate accounts for testing on a set far from the training one, i.e., the bootstrap replicate. This is because every sample case in the testing set has zero probability of belonging to the training dataset, i.e., very distant from the training dataset. This is a reason for why the LOOB is an upwardly biased estimator. Efron [8] showed roughly that:

$$\mathrm{E}_F\left[\mathrm{Err}_{\mathbf{t}} - \overline{\mathrm{Err}}_{\mathbf{t}}\right] \approx 0.632\,\mathrm{E}_F\left[\widehat{\mathrm{Err}}_{\mathbf{t}}^{(1)} - \overline{\mathrm{Err}}_{\mathbf{t}}\right]. \tag{36}$$

Substituting back in (33) gives the 0.632 estimator:

$$\widehat{\mathrm{Err}}_{\mathbf{t}}^{(0.632)} = 0.368\,\overline{\mathrm{Err}}_{\mathbf{t}} + 0.632\,\widehat{\mathrm{Err}}_{\mathbf{t}}^{(1)}. \tag{37}$$

The proof of the above results can be found in Efron [8] and Efron and Tibshirani [11, Sect. 6].

The motivation behind this estimator as stated earlier is to correct for the downward biased apparent error by adding a piece of the upward biased LOOB estimator. But an increase in variance should be expected as a result of adding this piece of the relatively variable apparent error. Moreover, this new estimator is no longer smooth since the apparent error itself is unsmooth.

3.3.4 The 0.632+ Bootstrap Estimator

The 0.632 estimator reduces the bias of the apparent error. But for over-trained classifiers, i.e., those whose apparent error tends to be zero, the 0.632 estimator is still downward biased. Breiman et al. [3] provided the example of an overfitted rule, like 1NN where the apparent error is zero. If, however, the class labels are assigned randomly to the predictors the true error rate will obviously be 0.5. But substituting in (37) gives an estimate of $0.632 \times 0.5 = 0.316$. To account for this bias for such over-fitted classifiers, Efron and Tibshirani [13] defined the *no-information error rate* γ by:

$$\gamma = \mathrm{E}_{0F_{ind}} L\left(y_0, \eta_{\mathbf{t}}(x_0)\right), \tag{38}$$

where F_{ind} means that x_0 and y_0 are distributed marginally as F but they are independent. Or said differently, the label is assigned randomly to the predictor. Then for a training sample $\mathbf{t}$, γ can be estimated by:

$$\hat{\gamma} = \frac{1}{n^2} \sum_{i=1}^{n} \sum_{j=1}^{n} L\left(y_i, \eta_{\mathbf{t}}(x_j)\right). \tag{39}$$

This means that the n predictors have been permuted with the n responses to produce n^2 non-informative cases. In the special case of binary classification, let $\hat{p}_1$ be the proportion of the response classified as belonging to class 1. Also, let $\hat{q}_1$ be the proportion of the responses classified as belonging to class 1. Then (39) reduces to:

$$\hat{\gamma} = \hat{p}_1(1 - \hat{q}_1) + (1 - \hat{p}_1)\hat{q}_1. \tag{40}$$

Also define the *relative overfitting rate*:

$$\hat{R} = \frac{\widehat{\text{Err}}_{\mathbf{t}}^{(1)} - \overline{\text{Err}}_{\mathbf{t}}}{\hat{\gamma} - \overline{\text{Err}}_{\mathbf{t}}}. \tag{41}$$

Efron and Tibshirani [13] showed that the bias of the 0.632 estimator for the case of over-fitted classifiers is alleviated by using a renormalized version of that estimator:

$$\widehat{\text{Err}}_{\mathbf{t}}^{(0.632+)} = (1 - \hat{w})\overline{\text{Err}}_{\mathbf{t}} + \hat{w}\widehat{\text{Err}}_{\mathbf{t}}^{(1)}, \tag{42a}$$

$$\hat{w} = \frac{0.632}{1 - 0.368\hat{R}}. \tag{42b}$$

It is useful to express the 0.632+ estimator in terms of its predecessor, the 0.632 estimator. Combining (37), (40), and (41) then substituting in (42a) yields:

$$\widehat{\text{Err}}_{\mathbf{t}}^{(0.632+)} = \widehat{\text{Err}}_{\mathbf{t}}^{(0.632)} + (\widehat{\text{Err}}_{\mathbf{t}}^{(1)} - \overline{\text{Err}}_{\mathbf{t}})\frac{0.368 \cdot 0.632 \cdot \hat{R}}{1 - 0.368\hat{R}}. \tag{43}$$

Efron and Tibshirani [13] consider the possibility that $\hat{R}$ lies out of the region [0, 1]. This leads to their proposal of defining:

$$\widehat{\text{Err}}_{\mathbf{t}}^{(1)\prime} = \min(\widehat{\text{Err}}_{\mathbf{t}}^{(1)}, \hat{\gamma}), \tag{44}$$

$$\hat{R}' = \begin{cases} (\widehat{\text{Err}}_{\mathbf{t}}^{(1)} - \overline{\text{Err}}_{\mathbf{t}})/(\hat{\gamma} - \overline{\text{Err}}_{\mathbf{t}}) & \overline{\text{Err}}_{\mathbf{t}} < \widehat{\text{Err}}_{\mathbf{t}}^{(1)} < \gamma \\ 0 & \text{otherwise} \end{cases}, \tag{45}$$

to obtain a modification to (43) that finally becomes:

$$\widehat{\text{Err}}_{\mathbf{t}}^{(0.632+)} = \widehat{\text{Err}}_{\mathbf{t}}^{(0.632)} + (\widehat{\text{Err}}_{\mathbf{t}}^{(1)\prime} - \overline{\text{Err}}_{\mathbf{t}})\frac{0.368 \cdot 0.632 \cdot \hat{R}'}{1 - 0.368\hat{R}'}. \tag{46}$$

3.4 Estimating the Standard Error of Error Rate Estimators

What have been reviewed above are several resampling methods: the CV, 0.632, and 0.632+ estimate the conditional error rate of a classification rule, conditional on that training dataset; and the LOOB estimates the mean error rate, where the expectation is taken over the population of training datasets. Regardless of what the estimator is designed to estimate, it is still a function of the current dataset $\mathbf{t}$, i.e., it is a random variable. If, e.g., the LOOB estimator $\widehat{\mathrm{Err}}_{\mathbf{t}}^{(1)}$ is considered, it estimates a constant real-valued parameter $\mathrm{E}_{0F}\mathrm{E}_F L(y_0, \eta_{\mathbf{t}}(x_0))$ with expectation taken over all the trainers and then over all the testers, respectively; this is the overall mean error rate. Yet, $\widehat{\mathrm{Err}}_{\mathbf{t}}^{(1)}$ is a random variable whose variability comes from the finite size of the available dataset. If the classifier is trained and tested on a very large number of observations, this would approximate training and testing on the entire population, and the variability would shrink to zero. This also applies for any performance measure other than the error rate. So, we are interested now in estimating $\mathrm{Var}_{\mathbf{t}}\widehat{\mathrm{Err}}_{\mathbf{t}}^{(1)}$, the variance of the estimator, not estimating $\mathrm{Var}_{\mathbf{t}}\mathrm{Err}_{\mathbf{t}}$, the variance of the true performance.

The next question then is, having estimated the mean performance of a classifier: what is the associated uncertainty of this estimate. Said differently: an estimate of the variance of this estimator be obtained from the same training dataset? Efron and Tibshirani [13] proposed the use of the IF method (Sect. 2.4), to estimate the uncertainty (variability) in $\widehat{\mathrm{Err}}_{\mathbf{t}}^{(1)}$. The reader is alerted that estimators that incorporate a piece of the apparent error are not suitable for the IF method. Such estimators are not smooth because the apparent error itself is not smooth.

By recalling the definitions of Sect. 2.4, $\widehat{\mathrm{Err}}_{\mathbf{t}}^{(1)}$ is now the statistic $s(\hat{F})$. To simplify notation, the error $L(y_i, \eta_{\mathbf{t}^{*b}}(x_i))$ may be denoted by L_i^b, and define the following notation:

$$l_{\cdot}^{b} = \frac{1}{n}\sum_{i=1}^{n} I_i^b L_i^b, \tag{47}$$

Also, define N_i^b to be the number of times the case t_i is included in the bootstrap b. Then, it has been proven in Efron and Tibshirani [12] that the IF of such an estimator is given by:

$$\left.\frac{\partial s(\hat{F}_{\varepsilon,i})}{\partial \varepsilon}\right|_{\varepsilon=0} = (2 + \frac{1}{n-1})(\hat{E}_i - \widehat{\mathrm{Err}}_{\mathbf{t}}^{(1)}) + \frac{n\sum_{b=1}^{B}(N_i^b - \bar{N}_i)I_i^b}{\sum_{b=1}^{B} I_i^b}. \tag{48}$$

Combining (78) and (48) gives an estimation to the uncertainty in $\widehat{\mathrm{Err}}_{\mathbf{t}}^{(1)}$.

4 Nonparametric Methods for Estimating the AUC of a Classification Rule

In the present section, we extend the study carried out in Efron [8], Efron and Tibshirani [13], and summarized in Sect. 3, to construct nonparametric estimators for the AUC (a two-sample statistic) analogue to those of the error rate (a one-sample statistic). Although some previous experimental comparative studies [26, 27, 32] were conducted to compare some of these resampling-based AUC estimators, in particular the 0.632 versions, there was no theoretical justification of using these estimators for the AUC. We provide here a full account of the different versions of bootstrap estimators reviewed in Sect. 3 and show how they can be formally extended to estimate the AUC.

4.1 Construction of Nonparametric Estimators for AUC

Before switching to the AUC, some more elaboration on Sect. 3 is needed. The SB estimator (29) can be rewritten as:

$$\widehat{\text{Err}}_{\mathbf{t}}^{SB} = \text{E}_{*}\text{E}_{\widehat{F}}\left[L(\eta_{\mathbf{t}^*}(x), y)|\mathbf{t}^*\right]. \tag{49}$$

Since there would be some observation overlap between $\mathbf{t}$ and $\mathbf{t}^*$, this approach suffers an obvious bias as was introduced in that section. This was the motivation behind interchanging the expectations and defining the LOOB (Sect. 3.3.1). Alternatively, we could have left the order of the expectation but with testing on only those observations in $\mathbf{t}$ that do not appear in the bootstrap replication $\mathbf{t}^*$, i.e., the distribution $\hat{F}^{(*)}$. The parenthesis notation $(*)$ refers to excluding from $\widehat{F}$, in the testing stage, the training cases $\mathbf{t}^*$ that were generated from the bootstrap replication. We call the resulting estimator $\widehat{\text{Err}}_{\mathbf{t}}^{(*)}$, which we define formally by:

$$\widehat{\text{Err}}_{\mathbf{t}}^{(*)} = \text{E}_{*}\text{E}_{\hat{F}^{(*)}}\left[L(\eta_{\mathbf{t}^*}(x), y)|\mathbf{t}^*\right] \tag{50}$$

We can give the inner expectation the notation $\text{Err}_{\mathbf{t}^{*b}}(\widehat{F}^{(*)})$, and rewrite the estimator as:

$$\widehat{\text{Err}}_{\mathbf{t}}^{(*)} = \text{E}_{*}\text{Err}_{\mathbf{t}^{*b}}(\widehat{F}^{(*)}) \tag{51a}$$

$$= \frac{1}{B}\sum_{b=1}^{B}\left[\sum_{i=1}^{N} I_i^b L(\eta_{\mathbf{t}^{*b}}(x_i), y_i) \Big/ \sum_{i'=1}^{N} I_{i'}^b\right], \tag{51b}$$

where the indicator I_i^b equals one if the observation t_i is excluded from the bootstrap replication $\mathbf{t}^{*b}$, and equals zero otherwise. The inner expectation in (50) is taken over those observations not included in the bootstrap replication $\mathbf{t}^*$, whereas the outer expectation is taken over all the bootstrap replications.

Analogously to Sect. 3, and to what has been introduced above, we can define several bootstrap estimators for the AUC. The start is the SB estimate, which can be defined as:

$$\widehat{\mathrm{AUC}}_{\mathbf{t}}^{SB} = \mathrm{E}_{*}\mathrm{AUC}_{\mathbf{t}^*}(\widehat{F}), \quad \widehat{F} \rightarrow \mathbf{t}^* \tag{52a}$$

$$= \mathrm{E}_{*}\left[\frac{1}{n_1 n_2}\sum_{j=1}^{n_2}\sum_{i=1}^{n_1}\psi(\hat{h}_{\mathbf{t}^*}(x_i), \hat{h}_{\mathbf{t}^*}(x_j))\right], \quad x_i \in \omega_1, x_j \in \omega_2. \tag{52b}$$

This averages the Mann-Whitney statistic over the bootstraps, where $\mathrm{AUC}_{\mathbf{t}^*}(\widehat{F})$ refers to the AUC obtained from training the classifier on the bootstrap replicate $\mathbf{t}^*$ and testing it on the empirical distribution $\widehat{F}$. In the approach used here, the bootstrap replicate $\mathbf{t}^*$ preserves the ratio between n_1 and n_2, which is called stratification. That is, the training sample $\mathbf{t}$ is treated as $\mathbf{t} = \mathbf{t}_1 \cup \mathbf{t}_2, \ \mathbf{t}_1 \in \omega_1, \ \mathbf{t}_2 \in \omega_2$; then n_1 cases are replicated from the first-class sample and n_2 cases are replicated from the second-class sample to produce $\mathbf{t}_1^*$ and $\mathbf{t}_2^*$ respectively, where $\mathbf{t}^* = \mathbf{t}_1^* \cup \mathbf{t}_2^*$. This was not needed when the performance measure was the error rate since it is a statistic that does not operate simultaneously on two different sets of observations as the Mann-Whitney statistic does (in U-statistic theory [25], error rate and Mann-Whitney are called one-sample and two-sample statistics respectively). The expectation (52a) is approximated by averaging over a finite number of bootstrap:

$$\widehat{\mathrm{AUC}}_{\mathbf{t}}^{SB} = \frac{1}{B}\sum_{b=1}^{B}\mathrm{AUC}_{\mathbf{t}^{*b}}(\widehat{F}), \tag{53}$$

The same motivation behind the estimator (32) can be applied here, i.e., testing only on those cases in $\mathbf{t}$ that are not included in the training dataset $\mathbf{t}^{*b}$, in order to reduce the bias. This can be carried out in (53) without interchanging the summation order. The new estimator is named $\widehat{\mathrm{AUC}}_{\mathbf{t}}^{(*)}$, where the parenthesis notation $(*)$ refers to the exclusion, in the testing stage, of the training cases that were generated from the bootstrap replication. Formally, we define this as:

$$\widehat{\mathrm{AUC}}_{\mathbf{t}}^{(*)} = \mathrm{E}_{*}\mathrm{AUC}_{\mathbf{t}^{*b}}(\widehat{F}^{(*)}) \tag{54a}$$

$$= \frac{1}{B}\sum_{b=1}^{B}\left[\sum_{j=1}^{n_2}\sum_{i=1}^{n_1}\psi(\hat{h}_{\mathbf{t}^*}(x_i), \hat{h}_{\mathbf{t}^*}(x_j))I_i^b I_j^b \Big/ \sum_{i'=1}^{n_1} I_{i'}^b \sum_{j'=1}^{n_2} I_{j'}^b\right]. \tag{54b}$$

The RB and 0.632 estimators can be introduced here in the same way it was used for the true error rate (Sect. 3.3.3) as:

$$\widehat{\mathrm{AUC}}_{\mathbf{t}}^{RB} = \overline{\mathrm{AUC}}_{\mathbf{t}} + \mathrm{E}_{*}\left[\mathrm{AUC}_{\mathbf{t}*}(\widehat{F}) - \overline{\mathrm{AUC}}_{\mathbf{t}*}\right]. \tag{55}$$

Then, if testing is carried out on cases excluded from the bootstraps, analogously to the 0.632 estimator of the error rate, this gives rise to the 0.632 estimator of the AUC:

$$\widehat{\mathrm{AUC}}_{\mathbf{t}}^{(0.632)} = 0.368\,\overline{\mathrm{AUC}_{\mathbf{t}}} + 0.632\,\widehat{\mathrm{AUC}}_{\mathbf{t}}^{(*)}. \tag{56}$$

It should be noted that this estimator is designed to estimate the true AUC for a classifier trained on the dataset **t** (the classifier performance conditional on the training dataset **t**). This is on contrary to the estimator (54) that estimates the mean performance of the classifier (this is the expectation over the training dataset population for the conditional performance).

The 0.632+ estimator $\widehat{\mathrm{AUC}}_{\mathbf{t}}^{(0.632+)}$ develops from $\widehat{\mathrm{AUC}}_{\mathbf{t}}^{(0.632)}$ in the same way as $\widehat{\mathrm{Err}}_{\mathbf{t}}^{(0.632+)}$ developed from $\widehat{\mathrm{Err}}_{\mathbf{t}}^{(0.632)}$ in Sect. 3.3.4. There are two modifications to the details. The first regards the *no-information error rate* γ; it can be proven that the *no-information* AUC is given by $\gamma_{\mathrm{AUC}} = 0.5$ (Lemma 2). The second regards the definitions (44), which should be modified to accommodate for the AUC. The new definitions are given by:

$$\widehat{\mathrm{AUC}}_{\mathbf{t}}^{(0.632+)} = \widehat{\mathrm{AUC}}_{\mathbf{t}}^{(0.632)} + (\widehat{\mathrm{AUC}}_{\mathbf{t}}^{(*)\prime} - \overline{\mathrm{AUC}_{\mathbf{t}}})\frac{0.368 \cdot 0.632 \cdot \hat{R}'}{1 - 0.368\hat{R}'}, \tag{57a}$$

$$\widehat{\mathrm{AUC}}_{\mathbf{t}}^{(*)\prime} = \max\big(\widehat{\mathrm{AUC}}_{\mathbf{t}}^{(*)}, \gamma_{\mathrm{AUC}}\big), \tag{57b}$$

$$\hat{R}' = \begin{cases} \frac{(\widehat{\mathrm{AUC}}_{\mathbf{t}}^{(*)} - \overline{\mathrm{AUC}_{\mathbf{t}}})}{(\gamma_{\mathrm{AUC}} - \overline{\mathrm{AUC}_{\mathbf{t}}})} & \text{if } \overline{\mathrm{AUC}_{\mathbf{t}}} > \widehat{\mathrm{AUC}}_{\mathbf{t}}^{(*)} > \gamma_{\mathrm{AUC}} \\ 0 & \text{otherwise} \end{cases}. \tag{57c}$$

To this end, we have constructed the AUC nonparametric estimators analogue to those of the error rate. Some of them, mainly the 0.632+ estimator, will have the least bias [13]. However, all of these estimators are not "smooth" and not eligible for the variance estimation via, e.g., the IF method (Sects. 2.4 and 3.4). The only estimator that may seem smooth, is the star versions $\widehat{\mathrm{Err}}_{\mathbf{t}}^{(*)}$ and $\widehat{\mathrm{AUC}}_{\mathbf{t}}^{(*)}$. However, the inner components $\mathrm{Err}_{\mathbf{t}^{*b}}(\widehat{F}^{(*)})$ and $\mathrm{AUC}_{\mathbf{t}^{*b}}(\widehat{F}^{(*)})$ are unsmooth themselves, because the classifier is trained on just one dataset. Applying the influence function enforces distributing the differential operator $\partial/\partial\varepsilon$, of the IF, over the summation to be encountered by these unsmooth components.

4.2 *The Leave-Pair-Out Boostrap (LPOB)* $\widehat{\mathrm{AUC}}^{(1,1)}$, *Its Smoothness and Variance Estimation*

The above discussion suggests introducing an analogue to $\widehat{\mathrm{Err}}_{\mathbf{t}}^{(1)}$ for measuring the performance in AUC. This estimator is motivated from (52a) the same way the estimator $\widehat{\mathrm{Err}}_{\mathbf{t}}^{(1)}$ was motivated from (31). The SB estimator (52a) can be rewritten as:

$$\widehat{\mathrm{AUC}}_{\mathbf{t}}^{SB} = \frac{1}{n_1 n_2} \sum_{j=1}^{n_2} \sum_{i=1}^{n_1} \mathrm{E}_* \psi(\hat{h}_{\mathbf{t}^*}(x_i), \hat{h}_{\mathbf{t}^*}(x_j)) \tag{58}$$

$$= \frac{1}{n_1 n_2} \sum_{j=1}^{n_2} \sum_{i=1}^{n_1} \sum_{b=1}^{B} \left[\psi(\hat{h}_{\mathbf{t}^{*b}}(x_i), \hat{h}_{\mathbf{t}^{*b}}(x_j)) \Big/ B \right]. \tag{59}$$

In words, the procedure is to select a pair (one observation from each class) and calculate for that pair the mean—over many bootstrap replications and training—of the Mann-Whitney kernel. Then, average over all possible pairs. This procedure will be optimistically biased because sometimes the testers will be the same as the trainers. To eliminate that bias, the inner bootstrap expectation should be taken only over those bootstrap replications that do not include the pair (t_i, t_j) in the training. Under that constraint, the estimator (58) becomes the leave-pair-out bootstrap (LPOB) estimator:

$$\widehat{\mathrm{AUC}}_{\mathbf{t}}^{(1,1)} = \frac{1}{n_1 n_2} \sum_{j=1}^{n_2} \sum_{i=1}^{n_1} \widehat{\mathrm{AUC}}_{i,j}, \tag{60a}$$

$$\widehat{\mathrm{AUC}}_{i,j} = \sum_{b=1}^{B} I_j^b I_i^b \psi(\hat{h}_{\mathbf{t}^{*b}}(x_i), \hat{h}_{\mathbf{t}^{*b}}(x_j)) \Big/ \sum_{b'=1}^{B} I_j^{b'} I_i^{b'}. \tag{60b}$$

The two estimators $\widehat{\mathrm{AUC}}_{\mathbf{t}}^{(*)}$ and $\widehat{\mathrm{AUC}}_{\mathbf{t}}^{(1,1)}$ produce very similar results; this is expected since they both estimate the same thing, i.e., the mean AUC. However, the inner component $\widehat{\mathrm{AUC}}_{i,j}$ of the estimator $\widehat{\mathrm{AUC}}_{\mathbf{t}}^{(1,1)}$ also enjoys the smoothness property of $\widehat{\mathrm{Err}}_{\mathbf{t}}^{(1)}$.

4.3 *Estimating the Standard Error of AUC Estimators*

The only smooth nonparametric estimator for the AUC so far is the LPOB estimator (60). Yousef et al. [33] discusses how to extend the approach of estimating the uncertainty in the error rate estimator using the IF method (Sect. 3.4) to estimate the uncertainty of this estimator, where interested readers may be referred to for all mathematical details and experimental results that show that the IF method provides almost unbiased estimation for the standard error of the LPOB estimator.

5 Illustrative Numerical Examples

5.1 *Error Rate Estimation*

Efron [8] and Efron and Tibshirani [13] provide comparisons of their proposed estimators (discussed in Sect. 3). They ran many simulations considering a variety of

Table 1 Average of RMS error of each estimator over 24 experiments run by Efron and Tibshirani [13]. The estimator $\widehat{\mathrm{Err}}_{\mathbf{t}}^{(1)}$ is the next to the estimator $\widehat{\mathrm{Err}}_{\mathbf{t}}^{(0.632+)}$ with only 2.5% increase in RMS

Estimator	Average RMS
$\mathrm{Err}_{\mathbf{t}}$	0
$\widehat{\mathrm{Err}}_{\mathbf{t}}^{(1)}$	0.083
$\widehat{\mathrm{Err}}_{\mathbf{t}}^{(0.632)}$	0.101
$\widehat{\mathrm{Err}}_{\mathbf{t}}^{(0.632+)}$	0.081
$\overline{\mathrm{Err}}_{\mathbf{t}}$	0.224

classifiers and data distributions, as well as real datasets. They assessed the estimators in terms of the RMS, the root of the experimental MSE:

$$\mathrm{MSE} = \mathrm{E}_{MC}(\widehat{\mathrm{Err}}_{\mathbf{t}} - \mathrm{Err}_{\mathbf{t}})^2 \tag{61a}$$

$$= \frac{1}{G}\sum_{g=1}^{G}(\widehat{\mathrm{Err}}_{\mathbf{t}_g} - \mathrm{Err}_{\mathbf{t}_g})^2, \tag{61b}$$

where $\widehat{\mathrm{Err}}_{\mathbf{t}_g}$ is the estimator (any estimator) conditional on a training dataset $\mathbf{t}_g$, and $\mathrm{Err}_{\mathbf{t}_g}$ is the true prediction error conditional on the same training dataset. The number of MC trials G in their experiments was 200. The following statement is quoted from Efron and Tibshirani [13]:

> The results vary considerably from experiment to experiment, but in terms of RMS error the 0.632+ rule is an overall winner.

This conclusion was without stating the criterion for deciding the *overall winner*. It was apparent from their results that the 0.632+ rule is the winner in terms of the bias—as was designed for. We calculated the average of the RMS of every estimator across all the 24 experiments they ran; Table 1 displays these averages. The estimators $\widehat{\mathrm{Err}}_{\mathbf{t}}^{(1)}$ and $\widehat{\mathrm{Err}}_{\mathbf{t}}^{(0.632+)}$ are quite comparable to each other with only 2.5% increase in the average RMS of the former. We will show below in Sect. 5.2 that the AUC estimators exhibit the same behavior but with magnified difference between the two estimators.

5.2 *AUC Estimation*

We carried out different experiments to compare the three bootstrap-based estimators $\widehat{\mathrm{AUC}}_{\mathbf{t}}^{(*)}$, $\widehat{\mathrm{AUC}}_{\mathbf{t}}^{(.632)}$, and $\widehat{\mathrm{AUC}}_{\mathbf{t}}^{(.632+)}$ considering different dimensionalities, different parameter values, and training set sizes. All experiments provided consistent and similar results. Here, in this section, we illustrate the results when the dimensionality $p = 5$, for multinormal 2-class data, with $\Sigma_1 = \Sigma_2 = \mathbf{I}$, $\mu_1 = \mathbf{0}$, $\mu_2 = c\mathbf{1}$, and c is an adjusting parameter to adjust the Mahalanobis distance

Table 2 Comparison of the different bootstrap-based estimators of the AUC. They are comparable to each other in the RMS sense, $\widehat{\mathrm{AUC}}_{\mathbf{t}}^{(.632+)}$ is almost unbiased, and all are weakly correlated with the true conditional performance $\mathrm{AUC}_{\mathbf{t}}$

Estimator	Mean	SD	RMS	RMS_{AM}	ρ	Size
$\mathrm{AUC}_{\mathbf{t}}$	0.6181	0.0434	0	0.0434	1.0000	
$\widehat{\mathrm{AUC}}_{\mathbf{t}}^{(*)}$	0.5914	0.0947	0.0973	0.0984	0.2553	
$\widehat{\mathrm{AUC}}_{\mathbf{t}}^{(0.632)}$	0.7012	0.0749	0.1128	0.1119	0.2559	20
$\widehat{\mathrm{AUC}}_{\mathbf{t}}^{(0.632+)}$	0.6431	0.0858	0.0906	0.0894	0.2218	
$\overline{\mathrm{AUC}}_{\mathbf{t}}$	0.8897	0.0475	0.2774	0.2757	0.2231	
$\mathrm{AUC}_{\mathbf{t}}$	0.6231	0.0410	0	0.0410	1.0000	
$\widehat{\mathrm{AUC}}_{\mathbf{t}}^{(*)}$	0.5945	0.0947	0.0956	0.0990	0.2993	
$\widehat{\mathrm{AUC}}_{\mathbf{t}}^{(0.632)}$	0.6991	0.0763	0.1066	0.1077	0.3070	22
$\widehat{\mathrm{AUC}}_{\mathbf{t}}^{(0.632+)}$	0.6459	0.0846	0.0863	0.0876	0.2726	
$\overline{\mathrm{AUC}}_{\mathbf{t}}$	0.8788	0.0499	0.2615	0.2606	0.2991	
$\mathrm{AUC}_{\mathbf{t}}$	0.6308	0.0400	0	0.0400	1.0000	
$\widehat{\mathrm{AUC}}_{\mathbf{t}}^{(*)}$	0.5991	0.0865	0.0897	0.0922	0.2946	
$\widehat{\mathrm{AUC}}_{\mathbf{t}}^{(0.632)}$	0.6971	0.0701	0.0961	0.0965	0.2997	25
$\widehat{\mathrm{AUC}}_{\mathbf{t}}^{(0.632+)}$	0.6442	0.0817	0.0815	0.0828	0.2758	
$\overline{\mathrm{AUC}}_{\mathbf{t}}$	0.8656	0.0471	0.2406	0.2395	0.2833	
$\mathrm{AUC}_{\mathbf{t}}$	0.6359	0.0358	0	0.0358	1.0000	
$\widehat{\mathrm{AUC}}_{\mathbf{t}}^{(*)}$	0.6035	0.0840	0.0874	0.0901	0.2904	
$\widehat{\mathrm{AUC}}_{\mathbf{t}}^{(0.632)}$	0.6962	0.0688	0.0906	0.0915	0.2934	28
$\widehat{\mathrm{AUC}}_{\mathbf{t}}^{(0.632+)}$	0.6479	0.0792	0.0785	0.0802	0.2719	
$\overline{\mathrm{AUC}}_{\mathbf{t}}$	0.8554	0.0472	0.2253	0.2246	0.2747	
$\mathrm{AUC}_{\mathbf{t}}$	0.6469	0.0343	0	0.0343	1.0000	
$\widehat{\mathrm{AUC}}_{\mathbf{t}}^{(*)}$	0.6170	0.0750	0.0792	0.0807	0.2746	
$\widehat{\mathrm{AUC}}_{\mathbf{t}}^{(0.632)}$	0.6997	0.0623	0.0818	0.0817	0.2722	33
$\widehat{\mathrm{AUC}}_{\mathbf{t}}^{(0.632+)}$	0.6553	0.0761	0.0752	0.0766	0.2656	
$\overline{\mathrm{AUC}}_{\mathbf{t}}$	0.8419	0.0439	0.2010	0.1999	0.2434	
$\mathrm{AUC}_{\mathbf{t}}$	0.6571	0.0308	0	0.0308	1.0000	
$\widehat{\mathrm{AUC}}_{\mathbf{t}}^{(*)}$	0.6244	0.0711	0.0753	0.0783	0.3185	
$\widehat{\mathrm{AUC}}_{\mathbf{t}}^{(.632)}$	0.6981	0.0598	0.0710	0.0725	0.3167	40
$\widehat{\mathrm{AUC}}_{\mathbf{t}}^{(.632+)}$	0.6595	0.0739	0.0707	0.0739	0.3092	
$\overline{\mathrm{AUC}}_{\mathbf{t}}$	0.8246	0.0431	0.1735	0.1730	0.2923	
$\mathrm{AUC}_{\mathbf{t}}$	0.6674	0.0271	0	0.0271	1.0000	
$\widehat{\mathrm{AUC}}_{\mathbf{t}}^{(*)}$	0.6357	0.0654	0.0690	0.0727	0.3534	
$\widehat{\mathrm{AUC}}_{\mathbf{t}}^{(.632)}$	0.6995	0.0556	0.0615	0.0642	0.3570	50
$\widehat{\mathrm{AUC}}_{\mathbf{t}}^{(.632+)}$	0.6685	0.0690	0.0646	0.0690	0.3522	
$\overline{\mathrm{AUC}}_{\mathbf{t}}$	0.8091	0.0406	0.1473	0.1474	0.3517	
$\mathrm{AUC}_{\mathbf{t}}$	0.6808	0.0217	0	0.0217	1.0000	
$\widehat{\mathrm{AUC}}_{\mathbf{t}}^{(*)}$	0.6533	0.0546	0.0602	0.0611	0.2451	
$\widehat{\mathrm{AUC}}_{\mathbf{t}}^{(.632)}$	0.7053	0.0471	0.0527	0.0531	0.2488	66
$\widehat{\mathrm{AUC}}_{\mathbf{t}}^{(.632+)}$	0.6840	0.0568	0.0556	0.0569	0.2477	
$\overline{\mathrm{AUC}}_{\mathbf{t}}$	0.7946	0.0355	0.1195	0.1192	0.2499	

(continued)

Table 2 (continued)

Estimator	Mean	SD	RMS	RMS_{AM}	ρ	Size
AUC_t	0.6965	0.0158	0	0.0158	1.0000	
$\widehat{AUC}_t^{(*)}$	0.6738	0.0454	0.0483	0.0507	0.3422	
$\widehat{AUC}_t^{(.632)}$	0.7119	0.0399	0.0405	0.0428	0.3492	100
$\widehat{AUC}_t^{(.632+)}$	0.7004	0.0452	0.0426	0.0453	0.3448	
$\overline{AUC}_t$	0.7772	0.0312	0.0860	0.0866	0.3596	
AUC_t	0.7141	0.0090	0	0.0090	1.0000	
$\widehat{AUC}_t^{(*)}$	0.6991	0.0298	0.0327	0.0334	0.2288	
$\widehat{AUC}_t^{(.632)}$	0.7205	0.0272	0.0273	0.0279	0.2291	200
$\widehat{AUC}_t^{(.632+)}$	0.7170	0.0285	0.0279	0.0286	0.2294	
$\overline{AUC}_t$	0.7573	0.0228	0.0487	0.0489	0.2277	

$\Delta = \left[(\mu_1 - \mu_2)'\Sigma^{-1}(\mu_1 - \mu_2)\right]^{1/2} = c^2 p$. We adjust c to keep a reasonable inter-class separation of $\Delta = 0.8$. When the classifier is trained, it will be tested on a pseudo-infinite test set, here 1000 cases per class, to obtain a very good approximation to the true AUC for the classifier trained on this very training dataset; this is called a single realization or a Monte-Carlo (MC) trial. Many realizations of the training datasets with same n are generated over MC simulation to study the mean and variance of the AUC for the Bayes classifier under this training set size. The number of MC trials is 1000 and the number of bootstraps is 100. It is apparent from Fig. 2 that the $\widehat{AUC}_t^{(*)}$ is downward biased. This is a natural opposite of the upward bias observed in Efron and Tibshirani [13] when the performance measure was the true error rate as a measure of incorrectness, by contrast with the true AUC

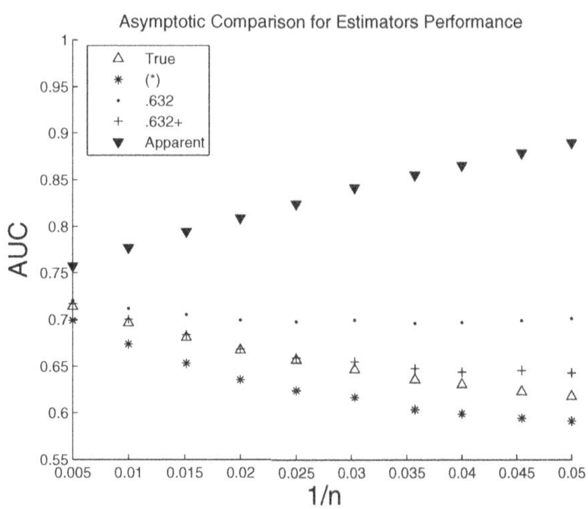

Fig. 2 Comparison of the three bootstrap estimators, $\widehat{AUC}_t^{(*)}$, $\widehat{AUC}_t^{(0.632)}$, and $\widehat{AUC}_t^{(0.632+)}$ for 5-feature predictor. The $\widehat{AUC}_t^{(*)}$ is downward biased, while the $\widehat{AUC}_t^{(0.632)}$ is an over correction for that bias. $\widehat{AUC}_t^{(0.632+)}$ is almost the unbiased version of the $\widehat{AUC}_t^{(0.632)}$. The figure first appeared in Yousef et al. [32]

Table 3 Average of RMS error of each estimator over the 10 experiments displayed in Table 2. The estimator $\widehat{\mathrm{AUC}}_{\mathbf{t}}^{(*)}$ is the next to $\widehat{\mathrm{AUC}}_{\mathbf{t}}^{(0.632+)}$ with only 9% increase in RMS

Estimator	Average RMS
$\mathrm{AUC}_{\mathbf{t}}$	0
$\widehat{\mathrm{AUC}}_{\mathbf{t}}^{(*)}$	0.07347
$\widehat{\mathrm{AUC}}_{\mathbf{t}}^{(0.632)}$	0.07409
$\widehat{\mathrm{AUC}}_{\mathbf{t}}^{(0.632+)}$	0.06735
$\overline{\mathrm{AUC}}_{\mathbf{t}}$	0.17808

as a measure of correctness. The $\widehat{\mathrm{AUC}}_{\mathbf{t}}^{(.632)}$ is designed as a correction for $\widehat{\mathrm{AUC}}_{\mathbf{t}}^{(*)}$; it appears in the figure to correct for that but with an over-shoot. The correct adjustment for the remaining bias is almost achieved by the estimator $\widehat{\mathrm{AUC}}_{\mathbf{t}}^{(.632+)}$. The $\widehat{\mathrm{AUC}}_{\mathbf{t}}^{(.632)}$ estimator can be seen as an attempt to balance between the two extreme biased estimators, $\widehat{\mathrm{AUC}}_{\mathbf{t}}^{(*)}$ and $\overline{\mathrm{AUC}}_{\mathbf{t}}$. However, it is expected that the component of $\overline{\mathrm{AUC}}_{\mathbf{t}}$ that is inherent in both $\widehat{\mathrm{AUC}}_{\mathbf{t}}^{(0.632+)}$ and $\widehat{\mathrm{AUC}}_{\mathbf{t}}^{(0.632)}$ increases the variance of these two estimators that my compensate for the decrease in the bias. Therefore, we assess all estimators in terms of the RMS, the root of the MSE defined in (61), and report the results in Table 2. In addition, we average the RMS of these estimators over the 10 experiments of Table 2 and list the average in Table 3. It is evident that the 0.632+ is slightly the overall winner with only 9% decrease in RMS if compared to the $\widehat{\mathrm{AUC}}_{\mathbf{t}}^{(*)}$ estimator. This almost agrees with the same result obtained for the error rate estimators and reported in Table 1.

In addition to the RMS, Table 2 compares the estimators in terms of the RMS around mean (RMS_{AM}): the root of the mean squared difference between an estimate and the mean performance (the mean over all possible training sets), instead of the conditional performance (conditional on a particular training set). The motivation behind that is explained next. The estimators $\widehat{\mathrm{AUC}}_{\mathbf{t}}^{(*)}$ and $\widehat{\mathrm{AUC}}_{\mathbf{t}}^{(1,1)}$ seem, at least from their formalization, to estimate the mean AUC of the classifier (this is the analogue of $\widehat{\mathrm{Err}}_{\mathbf{t}}^{(*)}$ and $\widehat{\mathrm{Err}}_{\mathbf{t}}^{(1)}$). However, the basic motivation for the $\widehat{\mathrm{AUC}}_{\mathbf{t}}^{(.632)}$ and $\widehat{\mathrm{AUC}}_{\mathbf{t}}^{(.632+)}$ is to estimate the AUC conditional on the given dataset $\mathbf{t}$ (this is the analogue of $\widehat{\mathrm{Err}}_{\mathbf{t}}^{(.632)}$ and $\widehat{\mathrm{Err}}_{\mathbf{t}}^{(.632+)}$). Nevertheless, as mentioned in Efron and Tibshirani [13] and detailed in Zhang [35] the CV, the basic ingredient of the bootstrap based estimators, is weakly correlated with the true performance on a sample by sample basis. This means that no estimator has a preference in estimating the conditional performance. Section 5.3 elaborates more on this phenomenon.

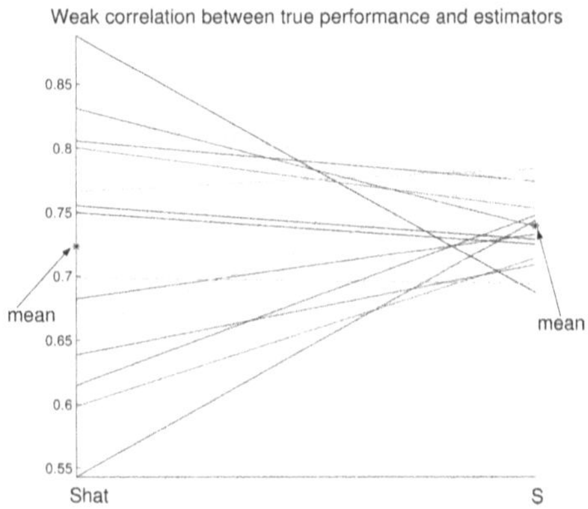

Fig. 3 The lack of correlation (or the weak correlation) between the bootstrap-based estimators and the true conditional performance. Every line connects the true performance of the classifier trained on a data set $\mathbf{t}_i$ and the estimated value. The figure represents 15 trials of the 1000 MC trials. Two nearby values of true performance may correspond to two widely separated estimates on different sides of the mean

5.3 Components of Variance and Weak Correlation

Many simulation results, e.g., Efron [8], Efron and Tibshirani [13], show that there is only a weak correlation between the CV estimator and the conditional true error rate $\text{Err}_{\mathbf{t}}$. This issue is discussed in mathematical detail in the excellent paper by Zhang [35], which therefore concludes that the CV estimator should not be used to estimate the true error rate of a classification rule conditional on a particular training data set. Other estimators discussed in the present article have this same attribute, since they have the same resampling ingredient of the CV estimator and "*we would guess, for any other estimate of conditional prediction error*" (Sect. 7.12, [20]). We provide our simple mathematical elaboration as follows. Denote the true performance of the classification rule conditional on the training set $\mathbf{t}$ (whether $\text{Err}_{\mathbf{t}}$, $\text{AUC}_{\mathbf{t}}$, or any other performance measure) by $S_{\mathbf{t}}$, the unconditional performance by $\text{E}_{\mathbf{t}} S_{\mathbf{t}}$, and an estimator of either of them by $\widehat{S}_{\mathbf{t}}$. For easier notation we can unambiguously drop the subscript $\mathbf{t}$ and decompose the MSE as

$$\text{MSE}(\widehat{S}, S) = \text{E}(\widehat{S} - S)^2 \tag{62a}$$

$$= \text{E}(\widehat{S} - \text{E}S)^2 + \text{Var}(S) - 2\text{Cov}(\widehat{S}, S). \tag{62b}$$

Then, by normalizing with the standard deviations we get:

$$\frac{\text{MSE}(\widehat{S}, S)}{\sigma_S \sigma_{\widehat{S}}} = \frac{\text{MSE}(\widehat{S}, \text{E}S)}{\sigma_S \sigma_{\widehat{S}}} + \frac{\sigma_S}{\sigma_{\widehat{S}}} - 2\rho_{\widehat{S}S}. \tag{63}$$

Table 4 Estimating the uncertainty in the estimator that estimates the difference in performance of two competing classifiers, the LDA and the QDA. The quantity M represents AUC_1 for LDA, AUC_2 for QDA, and Δ for the difference

Metric M	LDA	QDA	Δ
$\text{E}\, M_t$	0.7706	0.7163	0.0543
$\text{SD}\, M_t$	0.0313	0.0442	0.0343
$\text{E}\, \widehat{M}^{(1,1)}$	0.7437	0.6679	0.0758
$\text{SD}\, \widehat{M}^{(1,1)}$	0.0879	0.0944	0.0533
$\text{E}\, \widehat{\text{SD}}\, \widehat{M}^{(1,1)}$	0.0898	0.1003	0.0708
$\text{SD}\, \widehat{\text{SD}}\, \widehat{M}^{(1,1)}$	0.0192	0.0163	0.0228

This equation relates four crucial components to each other:

- $\text{MSE}(\widehat{S}, S)/\sigma_S\sigma_{\widehat{S}}$, the normalized MSE of $\widehat{S}$, if we see it as an estimator of the conditional performance S.
- $\text{MSE}(\widehat{S}, \text{E}S)/\sigma_S\sigma_{\widehat{S}}$, the normalized MSE of $\widehat{S}$, if we see it as an estimator of the expected performance $\text{E}S$ (and therefore called MSE around the mean).
- $\sigma_S/\sigma_{\widehat{S}}$, the standard deviation ratio between S and $\widehat{S}$.
- $\rho_{\widehat{S}S}$, the correlation coefficient between S and $\widehat{S}$.

From (63), an estimator $\widehat{S}$ is a good candidate to estimate S than $\text{E}S$ if its $\text{MSE}(\widehat{S}, S)$ is less than its $\text{MSE}(\widehat{S}, \text{E}S)$. Then, it is the responsibility of the correlation coefficient $\rho_{\widehat{S}S}$ to be high enough to cancel $\sigma_S/\sigma_{\widehat{S}}$ and a portion of $\text{MSE}(\widehat{S}, \text{E}S)$. Unfortunately, this is not the case as we illustrate experimentally in Table 2, which provides all quantities of the decomposition (63). It is obvious from the values that $\text{RMS}(\widehat{S}, S)$ and $\text{RMS}(\widehat{S}, \text{E}S)$ are very close to each other because the quantity $\sigma_S/\sigma_{\widehat{S}} - 2\rho_{\widehat{S}S} \simeq 0.413 - 2 \times 0.290 = -0.167$ (on average over the 10 experiments shown in the table). Moreover, in some cases, e.g., the first experiment, it goes as low as -0.052. The correlation between $\widehat{S}$ and S is weak to cast $\widehat{S}$ as an estimate to S, although it is designed to estimate it! For more illustration, Fig. 3 visualizes the components in Eq. (63) and the numbers in Table 2. This figure shows 15 realizations of the 1000 MC trials of the same experiment above. On the right, are the true values of S when trained on these different 15 training sets. On the left, are the corresponding 15 estimated values of $\widehat{S}$. The lines provide links between the true values and the corresponding estimates. This figure shows that two nearby true values of S are likely to have two widely separated estimated values $\widehat{S}$ on different sides of the mean. This visually illustrates the lack of correlation (or the weak correlation) between the estimators and the true conditional performance.

5.4 *Two Competing Classifiers*

If the assessment problem is how to compare two classifiers, rather than the individual performance, then the measure to be used is either the conditional difference

$$\Delta_{\mathrm{t}} = \mathrm{AUC}_{1_{\mathrm{t}}} - \mathrm{AUC}_{2_{\mathrm{t}}}, \tag{64}$$

or the mean, unconditional, difference

$$\Delta = \mathrm{E}\,\Delta_{\mathrm{t}} = \mathrm{E}\left[\mathrm{AUC}_{1_{\mathrm{t}}} - \mathrm{AUC}_{2_{\mathrm{t}}}\right], \tag{65}$$

where, we defined them for the AUC just for illustration with immediate identical treatment for other measures. Then it is obvious that there is nothing new in the estimation task, i.e., it is merely the difference of the performance estimate of each classifier, i.e.,

$$\widehat{\Delta} = \mathrm{E}\,\widehat{\mathrm{AUC}}_{1_{\mathrm{t}}} - \mathrm{E}\,\widehat{\mathrm{AUC}}_{2_{\mathrm{t}}}, \tag{66}$$

where each of the two estimators in (66) is obtained by any estimator. A natural candidate, from the point of view of the present chapter is the LPOB estimator $\widehat{\mathrm{AUC}}^{(1,1)}$—because of both the smoothness and weak correlation issues discussed so far.

Then, how to estimate the uncertainty (variance) of $\widehat{\Delta}$. This is very similar to estimating the variance in $\widehat{\mathrm{E}\,\mathrm{AUC}_{\mathrm{t}}}$. There is nothing new in estimating $\mathrm{Var}\,\widehat{\Delta}$. It is obtained by replacing $\widehat{\mathrm{AUC}}^{(1,1)}$, in Yousef et al. [33], by the statistic $\widehat{\Delta}$ in (66). For demonstration, typical values are given in Table 4, for comparing the linear and quadratic discriminants, where the training set size per class is 20 and number of features is 4.

6 Discussion and Conclusion

In this chapter, the very important topic of the assessment of ML algorithms is reviewed, with an emphasis on the nonparametric assessment of classification rules. The topic is quite important to many fields and applications, in particular cyberphysical security, where ML algorithms are almost ubiquitous. We started with reviewing the basic nonparametric methods for estimating the bias and variance of a statistic. Then, we reviewed the basic resampling-based methods for estimating the error rate of a classification rule. Departing from that, we extended these estimators from estimating the error rate (a one-sample statistic) to estimating the AUC (a two-sample statistic). This extension is theoretically justified, and not just an ad hoc application. Among these estimators, we identified those that are smooth and eligible for estimating their standard error using the IF method.

It was interesting to see, through the whole chapter, the connection among different resampling-based estimators. It is worth mentioning that, in addition to the conventional K-fold CV, there are other versions and variants, which are usually used in an ad hoc way by many practitioners. The formalization of these versions and variants, and the mathematical connection among them, along with their connection to the bootstrap-based estimators, all can be established in the same spirit and approach followed in the present chapter. However, many of them are unsmooth except possibly the repeated CV, which is partially smooth and suitable for the IF method [30, 31].

With this rich variety of estimators, a practitioner may legitimately wonder about the "optimal" estimator (in terms of any optimality criterion) that should be systematically used. There are three aspects, on which we can base our comparison: accuracy, uncertainty estimation, and computational efficiency.

In terms of accuracy, it is surprising to know that, from the few number of comparative studies available in the literature, there is no overall winner among these estimators. All of them have comparable accuracy, measured in terms of RMS, with a little superiority of the 0.632+ bootstrap estimator. In addition, and most importantly, all estimators have a weak correlation with the true conditional performance (e.g., $\mathrm{Err}_{\mathbf{t}}$, the conditional error rate, or $\mathrm{AUC}_{\mathbf{t}}$, the conditional AUC), a phenomenon that allows them to be eligible only for estimating the mean true performance (e.g., $\mathrm{E}_{\mathbf{t}}\mathrm{Err}_{\mathbf{t}}$ or $\mathrm{E}_{\mathbf{t}}\mathrm{AUC}_{\mathbf{t}}$), where the mean is taken over the population of training datasets as explained through the chapter. Said differently, the performance estimation that a practitioner obtains using, e.g., the CV, is not an estimation of the performance of this very trained ML algorithm; rather, it is an estimation of the mean performance of this algorithm had we trained it on all possible training datasets of the same size! We quote from [20, Sect. 7.12]:

> *This phenomenon also occurs for bootstrap estimates of error, and we would guess, for any other estimate of conditional prediction error.*

In terms of the variance estimation of these estimators (not the estimation of the variance of the algorithm itself), only a few of them are smooth and candidates for a sophisticated method like the IF. The ordinary K-fold CV is not among those! Rather, only the computationally expensive version of it, the repeated CV, is partially smooth as mentioned above.

It terms of the computational aspects, the bootstrap-based estimators are computationally expensive. If compared to the conventional K-fold CV, which requires only K iterations of both training and testing, the former require hundreds of bootstrap replications. Because the majority of recent ML applications involve both massive datasets and complex algorithms, including DNN that is very computationally expensive, it is obvious that the CV may be more practical than the bootstrap-based estimators. However, for some other fields, e.g., cyberphysical security, many applications produce tabular (structured) data. Tabular data are more suitable for the traditional and less computationally expensive ML algorithms. Therefore, serious practitioners in these fields and applications may need to keep all of these estimators in their toolbox. Moreover, it is quite prudent to see a future benchmark that compiles these estimators, along with different datasets from a wide range of applications, in a single comprehensive comparative study.

Acknowledgements The author is grateful to the U.S. Food and Drug Administration (FDA) for funding a very early stage of this chapter, and to Dr. Kyle Myers for her support. In his memorial, special thanks and gratitude to Dr. Robert F. Wagner, the supervisor and the teacher, or Bob Wagner, the big brother and friend. He reviewed a very early version of this chapter before he passed away.

7 Appendix

7.1 Proofs

Lemma 1 *The maximum likelihood estimation (MLE) for the probability mass function under nonparametric distribution, given a sample of n observations, is given by:*

$$\hat{F}: \; mass \; \frac{1}{n} \; on \; t_i, \quad i = 1, \ldots, n. \tag{67}$$

Proof The proof is carried out by maximizing the likelihood function $l(f) = \prod_{i=1}^{n} p_i$, which can be rewritten under the constraint $\sum_i p_i = 1$, using a Lagrange's multiplier, as:

$$l(f) = \prod_{i=1}^{n} p_i + \lambda \left(\sum_{i=1}^{n} p_i - 1 \right). \tag{68}$$

The likelihood (68) is maximized by taking the first derivative and setting it to zero to obtain:

$$\frac{\partial l(f)}{\partial p_j} = \prod_{i \neq j} p_i + \lambda \overset{set}{=} 0, \quad j = 1, \ldots, n. \tag{69}$$

These n equations along with the constraint $\sum_i p_i = 1$ can be solved straightforwardly to give $\hat{p}_i = \frac{1}{n}$, $i = 1, \ldots, n$, which completes the proof. □

Lemma 2 *The no-information* AUC *is given by* $\gamma_{\mathrm{AUC}} = 0.5$.

Proof γ_{AUC}, an analogue to the *no-information error rate* γ, is given by (2a) but with TPF and FPF given under the *no-information* distribution E_{0F} (see Sect. 3.3.4). Therefore, assume that there are n_1 and n_2 observations from class ω_1 and ω_2, respectively. Assume also for a fixed threshold th the two quantities that define the error rate are TPF and FPF. Also, assume that the sample observations are tested by the classifier and each sample has been assigned a decision value (score). Under the *no-information* distribution, consider the following. For every decision value $h_t(x_i)$ assigned for the observation $t_i = (x_i, y_i)$, create new $n_1 + n_2 - 1$ observations; all of them have the same decision value $h_{\mathbf{t}}(x_i)$, while their responses are equal to the responses of the rest $n_1 + n_2 - 1$ observations $t_j,\ j \neq i$. Under this new sample that consists of $(n_1 + n_2)^2$ observations, it is quite easy to see that the new TPF and FPF for the same threshold th are given by $\text{FPF}_{0\widehat{F},th} = \text{TPF}_{0\widehat{F},th} = (\text{TPF} \cdot n_1 + \text{FPF} \cdot n_2)/(n_1 + n_2)$. This means that the ROC curve under the *no-information* rate is a straight line with slope equal to one; this directly gives $\gamma_{\text{AUC}} = 0.5$.

7.2 More on Influence Function (IF)

Assume that there is a distribution G near to the distribution F; then under some regularity conditions(see, e.g., [21], Chap. 2) a functional s can be approximated as:

$$s(G) \approx s(F) + \int IC_{s,F}(x)\, dG(x). \tag{70}$$

The residual error can be neglected since it is of a small order in probability. Some properties of (70) are:

$$\int IC_{T,F}(x)\, dF(x) = 0, \tag{71}$$

and the asymptotic variance of $s(F)$ under F, following from (71), is given by:

$$\text{Var}_F s(F) \simeq \int \left[IC_{T,F}(x)\right]^2\, dF(x), \tag{72}$$

which can be considered as an approximation to the variance under a distribution G near to F. Now, assume that the functional s is a functional statistic in the dataset $\mathbf{x} = \{x_i : x_i \sim F,\ i = 1, \ldots, n\}$. In that case the influence curve (23) is defined for each sample case x_i, under the true distribution F as:

$$U_i(s, F) = \lim_{\varepsilon \to 0} \frac{s(F_{\varepsilon,i}) - s(F)}{\varepsilon} = \left.\frac{\partial s(F_{\varepsilon,i})}{\partial \varepsilon}\right|_{\varepsilon=0}, \tag{73}$$

where $F_{\varepsilon,i}$ is the distribution under the perturbation at observation x_i. Equation (73) is called the IF. If the distribution F is not known, the MLE $\hat{F}$ of the distribution F is given by (3), and as an approximation $\hat{F}$ may substitute for F in (73). The result may then be called the empirical IF [24], or infinitesimal jackknife [22]. In such an approximation, the perturbation defined in (22) can be rewritten as:

$$\hat{F}_{\varepsilon,i} = (1 - \varepsilon)\hat{F} + \varepsilon\delta_{x_i},\quad x_i \in \mathbf{x},\quad i = 1, \ldots, n. \tag{74}$$

This kind of perturbation is illustrated in Fig. 4.

It will often be useful to write the probability mass function of (74) as:

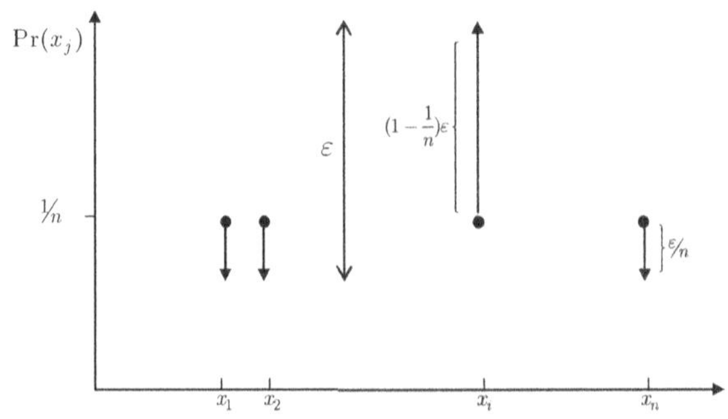

Fig. 4 The new probability masses for the dataset **x** under a perturbation at sample case x_i obtained by letting the new probability, at x_i exceed the new probability at any other case x_i by, ε

$$\hat{f}_{\varepsilon,i}(x_j) = \begin{cases} \frac{1-\varepsilon}{n} + \varepsilon & j = i \\ \frac{1-\varepsilon}{n} & j \neq i \end{cases}. \tag{75}$$

A very interesting case arises from (75) if $-1/(n+1)$ is substituted for ε. In this case the new probability mass assigned to the point $x_{j=i}$ in (75) will be zero. This value of ε simply generates the jackknife estimate discussed in Sect. 2.2, where the whole observation is removed from the dataset.

Substituting $\hat{F}$ for G in (70) and combining the result with (73) gives the IF approximation for any functional statistic under the empirical distribution $\hat{F}$. The result is:

$$s(\hat{F}) = s(F) + \frac{1}{n}\sum_{i=1}^{n} U_i(s, F) + O_p(n^{-1}) \tag{76a}$$

$$\approx s(F) + \frac{1}{n}\sum_{i=1}^{n} U_i(s, F). \tag{76b}$$

The term $O_p(n^{-1})$ reads "big-O of order $1/n$ in probability". In general, $U_n = O_p(d_n)$ if U_n/d_n is bounded in probability, i.e., $\Pr\{|U_n|/d_n < k_\varepsilon\} > 1 - \varepsilon \; \forall \, \varepsilon > 0$. This concept can be found in [1, Chap. 2]. Then the asymptotic variance expressed in (72) can be given for $s(F)$ by:

$$\mathrm{Var}_F s = \frac{1}{n} \mathrm{E}_F U^2(x_i, F), \tag{77}$$

which can be approximated under the empirical distribution $\hat{F}$ to give the nonparametric estimate of the variance for a statistic s by:

$$\widehat{\mathrm{Var}}_{\hat{F}} s = \frac{1}{n^2} \sum_{i=1}^{n} U_i^2(x_i, \hat{F}). \tag{78}$$

7.3 ML in Other Fields

In this section we provide very brief miscellanea from other fields for the reader to see a bigger picture of this chapter. As already was mentioned, ML is crucial to many applications. For example, in the medical imaging field, a tumor on a mammogram must be classified as malignant or benign. This is an example of prediction, regardless of whether it is done by a radiologist or by a computer aided detection (CAD) software. In either case, the prediction is done based on learning from previous mammograms. The features, i.e., predictors, in this case may be the size of the tumor, its density, various shape parameters, etc. The output, i.e., response, is categorical and belongs to the set: $\mathcal{G} = \{benign, \ malignant\}$. There are so many such examples in biology and medicine that it is almost a field unto itself, i.e., biostatistics. The task may be diagnostic as in the mammographic example, or prognostic where, for example, one estimates the probability of occurrence of a second heart attack for a particular patient who has had a previous one. All of these examples involve a prediction step based on previous learning. A wide range of commercial and military applications arises in the field of satellite imaging. Predictors in this case can be measures from the image spectrum, while the response can be the type of land, crop, or vegetation of which the image was taken.

Some expressions and terminology of ML belong to some fields and applications more than the others. E.g., it is conventional in medical imaging to refer to e_1 as the false negative fraction (FNF), and e_2 as the false positive fraction (FPF). This is because diseased patients typically have a higher output value for a test than non-diseased patients. For example, a patient belonging to class 1 whose test output value is less than the threshold setting for the test will be called "test negative", while the patient is in fact in the diseased class. This is a false negative decision; hence the name FNF. The situation is reversed for the other error component.

The importance of the AUC is natural and unquestionable in some applications than others. The equivalence of the area under the empirical ROC and the Mann-Whitney-Wilcoxon statistic is the basis of its use in the assessment of diagnostic tests; see Hanley and McNeil [19]. Swets [29] has recommended it as a natural summary measure of detection accuracy on the basis of signal-detection theory. Applications of this measure are widespread in the literature on both human diagnosis and computer-aided diagnosis, in medical imaging [23]. In the field of machine learning, Bradley [2] has recommended it as the preferred summary measure of accuracy when a single number is desired. These references also provide general background and access to the large literature on the subject.

Even the mistakes committed by some practitioners are obvious in some fields more than others. E.g., in DNA microarrays, these mistakes are fatal and produce very fragile results. This is because of the very high dimensionality of the problem with respect to the amount of available dataset. A more elaborate assessment phase should follow the design and construction phase in such ill-posed applications.

References

1. Barndorff-Nielsen OE, Cox DR (1989) Asymptotic techniques for use in statistics. Chapman and Hall, New York
2. Bradley AP (1997) The use of the area under the ROC curve in the evaluation of machine learning algorithms. Pattern Recogn 30(7):1145
3. Breiman L, Friedman J, Olshen R, Stone C (1984) Classification and regression trees. Wadsworth International Group, Belmont
4. Chen W, Gallas BD, Yousef WA (2012) Classifier variability: accounting for training and testing. Pattern Recogn 45(7):2661–2671
5. Efron B (1979) Bootstrap methods: another look at the Jackknife. Ann Stat 7(1):1–26
6. Efron B (1981) Nonparametric estimates of standard error: the Jackknife, the bootstrap and other methods. Biometrika 68(3):589–599
7. Efron B (1982) The Jackknife, the bootstrap, and other resampling plans. Society for Industrial and Applied Mathematics, Philadelphia
8. Efron B (1983) Estimating the error rate of a prediction rule: improvement on cross-validation. J Am Stat Assoc 78(382):316–331
9. Efron B (1986) How biased is the apparent error rate of a prediction rule? J Am Stat Assoc 81(394):461–470
10. Efron B, Stein C (1981) The Jackknife estimate of variance. Ann Stat 9(3):586–596
11. Efron B, Tibshirani R (1993) An introduction to the bootstrap. Chapman and Hall, New York
12. Efron B, Tibshirani R (1995) Cross validation and the bootstrap: estimating the error rate of a prediction rule. Technical report 176, Stanford University, Department of Statistics
13. Efron B, Tibshirani R (1997) Improvements on cross-validation: the .632+ Bootstrap method. J Am Stat Assoc 92(438):548–560
14. Fukunaga K (1990) Introduction to statistical pattern recognition, 2nd edn. Academic Press, Boston
15. Hájek J, Šidák Z, Sen PK (1999) Theory of rank tests, 2nd edn. Academic Press, San Diego
16. Hampel FR (1974) The influence curve and its role in robust estimation. J Am Stat Assoc 69(346):383–393
17. Hampel FR (1986) Robust statistics?: the approach based on influence functions. Wiley, New York
18. Hanley JA (1989) Receiver operating characteristic (ROC) methodology: the state of the art. Crit Rev Diagn Imaging 29(3):307–335
19. Hanley JA, McNeil BJ (1982) The meaning and use of the area under a receiver operating characteristic (ROC) curve. Radiology 143(1):29–36
20. Hastie T, Tibshirani R, Friedman JH (2009) The elements of statistical learning: data mining, inference, and prediction, 2nd edn. Springer, New York
21. Huber PJ (1996) Robust statistical procedures, 2nd edn. Society for Industrial and Applied Mathematics, Philadelphia
22. Jaeckel L (1972) The infinitesimal jackknife. Memorandum, MM 72-1215-11, Bell Lab Murray Hill
23. Jiang Y, Nishikawa RM, Schmidt RA, Metz CE, Giger ML, Doi K (1999) Improving breast cancer diagnosis with computer-aided diagnosis. Acad Radiol 6(1):22–33
24. Mallows C (1974) On some topics in robustness. Memorandum, MM 72-1215-11, Bell Lab Murray Hill, NJ
25. Randles RH, Wolfe DA (1979) Introduction to the theory of nonparametric statistics. Wiley, New York
26. Sahiner B, Chan HP, Petrick N, Hadjiiski L, Paquerault S, Gurcan MN (2001) Resampling schemes for estimating the accuracy of a classifier designed with a limited data set. In: Medical image perception conference IX, airlie conference Center, Warrenton VA, 20–23
27. Sahiner B, Chan HP, Hadjiiski L (2008) Classifier performance prediction for computer-aided diagnosis using a limited dataset. Med Phys 35(4):1559

28. Stone M (1974) Cross-validatory choice and assessment of statistical predictions. J Roy Stat Soc: Ser B (Methodol) 36(2):111–147
29. Swets JA (1986) Indices of discrimination or diagnostic accuracy: their ROCs and implied models. Psychol Bull 99:100–117
30. Yousef WA (2019) A leisurely look at versions and variants of the cross validation estimator. arXiv preprint arXiv:1907.13413
31. Yousef WA (2021) Estimating the standard error of cross-validation-based estimators of classifier performance. Pattern Recogn Lett 146:115–145
32. Yousef WA, Wagner RF, Loew MH (2004) Comparison of non-parametric methods for assessing classifier performance in terms of ROC parameters. In: Proceedings of 33rd applied imagery pattern recognition workshop, 2004. IEEE Computer Society, pp 190–195
33. Yousef WA, Wagner RF, Loew MH (2005) Estimating the uncertainty in the estimated mean area under the ROC curve of a classifier. Pattern Recogn Lett 26(16):2600–2610
34. Yousef WA, Wagner RF, Loew MH (2006) Assessing classifiers from two independent data sets using ROC analysis: a nonparametric approach. IEEE Trans Pattern Anal Mach Intell 28(11):1809–1817
35. Zhang P (1995) Assessing prediction error in nonparametric regression. Scand J Stat 22(1):83–94

A Collection of Datasets for Intrusion Detection in MIL-STD-1553 Platforms

Hadeer Ahmed, Issa Traore, Paulo Quinan, Karim Ganame, and Oussama Boudar

Abstract MIL-STD-1553 is a military standard communication protocol that has been around for over four decades and is central to the operation of a wide range of defense platforms. At its inception, the standard was conceived with a focus only on reliability and fault tolerance, with no attention paid to security concerns. However, it has been shown in the last few years that modern defense platforms are increasingly the target of cyber-attacks from both state and non-state actors. In such a context, MIL-STD-1553 data buses represent prime conduits for compromising defense platforms that rely on them for communications. This chapter explores a range of cyberattacks against MIL-STD-1553 data buses and present a collection of datasets that were generated by executing a selected attack scenarios in a testbed environment. It is expected that the proposed datasets can be used toward designing and evaluating intrusion detection systems for MIL-STD-153 avionic platforms.

Keywords Collection · Dataset · Intrusion detection system · MIL-STD-1553 · Defense platform · Avionic platform · Cyberattack · Protocol · Unsupervised detection · Vulnerabilities · Attack scenario · Unsupervised machine learning

H. Ahmed (✉) · I. Traore · P. Quinan
Department of Electrical and Computer Engineering, University of Victoria, Victoria, BC, Canada
e-mail: hsahmed@uvic.ca

I. Traore
e-mail: itraore@ece.uvic.ca

P. Quinan
e-mail: quinan@uvic.ca

K. Ganame · O. Boudar
Efficient Protections Inc., Ottawa, QC, Canada
e-mail: ganame@streamscan.io

O. Boudar
e-mail: oussama.boudar@streamscan.io

I. Traore et al. (eds.), *Artificial Intelligence for Cyber-Physical Systems Hardening*, Engineering Cyber-Physical Systems and Critical Infrastructures 2,
https://doi.org/10.1007/978-3-031-16237-4_4

1 Introduction

This MIL-STD-1553 is a standard communication bus that interconnects different terminals and devices involved in military vehicles. It is a serial communication bus which uses asynchronous Time Division Multiplexing (TDM), which means that each device maintains and uses its own clock for transmission.

It is central to the operation of a broad range of defense platforms deployed on various major military aircraft [1]. However, it was designed at a time when cybersecurity was an afterthought. Indeed, while MIL-STD-1553 was designed with built-in reliability and fault tolerance, it has no inherent cybersecurity protections. There are many glaring security flaws in the protocol which could be exploited by determined and skilled actors, which is the hallmark of state-actors who are known to be constantly probing and threatening national defense posture [2, 3].

As highlighted in a master's thesis conducted by Blaine Losier at the Royal Military College (RMC) of Canada, classified research, recently performed, has exposed a number of vulnerabilities in the MIL-STD-1553 protocol [4]. For instance, in [5], the authors demonstrated the feasibility of cyber-attacks against MIL-STD-1553 buses used for communications between subsystems in a satellite.

Because a complete redesign of the protocol would require a major overhaul of the thousands of systems that depend on the bus, a solution that has been advocated in the recent literature is to mitigate the security flaws by deploying adequate intrusion detection and response systems [6]. An intrusion detection system (IDS) can fit neatly in the current MIL-STD-1553 architecture through the so-called Bus Monitor (BM), which provides a standard placeholder for monitoring devices hooked on the bus. This would represent a cost-effective alternative to a major overhaul of the protocol and its dependent systems.

Our goal is to develop a new anomaly detection model for MIL-STD-1553 that uses unsupervised machine learning models. To our knowledge, all the existing relevant IDS proposals focus on supervised models, which may not be effective in the context of MIL-STD-1553 systems. Supervised anomaly detection requires a labelled dataset to train the corresponding model. In contrast, unsupervised techniques do not depend on labelled data for detection. We believe that unsupervised detection is more suitable for MIL-STD-1553 platforms due to the limited availability of prior knowledge on the attack methods targeted at these platforms due to their novelty and the restricted classification applied for information related to military cyber incidents. A system based on this kind of anomaly detection techniques will be able to detect a much broader range of anomalies, including novel and unseen ones. The main challenge in using unsupervised machine learning for detecting anomalies is the need to design an effective and efficient mechanism for determining what is normal for the time series being monitored. Another major challenge in designing an anomaly detection system is the availability of adequate datasets. To our knowledge, only one relevant dataset has been released to the public [7].

In this chapter, we explore different attack scenarios and use such knowledge to generate new datasets through simulation.

The rest of the chapter is structured as follows.

Section 2 presents the key concepts and requirements that underlie the normal operation of 1553 standard. These provide a foundation to define baseline model for the normal operation and behavior of MIL-STD-1553 systems.

Section 3 presents different potential attack vectors against MIL-STD-1553. While such enumeration of attack methods is far from being exhaustive, it helps sketch different attack scenarios which can be used in evaluating MIL-STD-1553 IDS.

Section 4 presents the simulation environment and the procedures and scenarios used to generate our dataset. Section 5 makes concluding remarks.

2 Mil-STD-1553 Baseline

The MIL-STD-1553 architecture consists of 5 generic components: bus controller (BC), bus monitor (BM), remote terminals (RTs), couplers and the bus itself [1]. A coupler isolates connected components from one another and in doing so it helps shield the bus from damage resulting from component malfunction.

The architecture is based on master–slave topology structured around a dual-redundant serial communication bus. Only one channel transmits data over the bus at a time, while the other channel serves as backup. If the transmission of a message on the primary bus failed, it will be retransmitted on the backup channel.

2.1 Major Components

Bus controller (*BC*): The communication between the components that are connected to the bus is orchestrated in a centralized way around the BC.

The BC manages communications between the bus components using command/response messages and following a strict predefined order and timing.

The BC is the only component allowed to initiate and mediate the communications between the components that are connected to the bus.

While several connected components may have BC capability, only one active BC is allowed at a time.

Remote Terminal (*RT*): An RT is any component that does not operate as BC or as BM. RT is a component whose core function is to transmit a message when instructed to do so by the BC. It cannot initiate a communication on its own. The BC can handle up to 31 connected RTs.

A key module of the RT is the subsystem, which represents the underlying computational unit responsible for the computations and data processing involved in executing the functionality performed by the RT.

Bus Monitor (*BM*): As the name indicates, the BM is responsible for observing the state and operation of the system. The BM is a passive component in the sense that it does not send any message and does not interfere with the operation and activity of the bus. Furthermore, it does not have any assigned Terminal Address (TA).

However, the BM can also have RT's capabilities, in addition to the core monitoring feature. BM can also serve as a backup BC.

2.2 *Bus Communication*

Communications between components are done by exchanging messages, which go through the BC. The data structure used in these communications are called "word". A message consists of a sequence of words, always starting with a command word. Each word is 20 bits long, including 3 synchronization bits at the beginning and a parity bit at the end.

Messages can be sent periodically, at fixed time intervals, or asynchronously based on some event-based triggers.

2.2.1 Frames

Major and minor frames are structures used by the BC to schedule the transmission of messages, whether periodically or asynchronously.[1]

Major frame: a predefined time interval over which all periodic messages are transmitted at least once. Each message is repeated periodically at a rate between 50 and 0.5 times per second. The period usually corresponds to the time of the message with lowest frequency in the schedule.

Although aperiodic (i.e., asynchronous) messages are dependent on specific (fixed) time interval, they are also part of a major frame and as such, a fixed time slot is defined in the frame for them accordingly.

Minor frame: consists of a sequence of messages with predefined inter message transmission gap times. The period of the minor frame typically matches the period of the highest frequency message on the bus schedule.

The period of a minor frame must always be less than the minimum period among aperiodic messages. Failure to do so will create frame overflow, whereby the update timer for sending some information may trigger before the end of the minor frame.

2.2.2 Words

There are 3 types of words: command, data and status, which are outlined in Table 1.

[1] The bus message schedule for an emulated BC is also called Bus list, which consists of a series of minor frames of equal duration and may also include a set of aperiodic messages.

Table 1 Mil-std-1553 words

	Fields	Position
Command word	Sync	1–3
	RT address	4–8
	T/R	9
	Sub-address/mode	10–14
	Data count/mode code	15–19
	Parity	20
Data word	Sync	1–3
	Data	4–19
	Parity	20
Status word	Sync	1–3
	RT address	4–8
	Message error	9
	Incrementation	10
	Service request	11
	Reserved	12–14
	Broadcast command received	15
	Busy	16
	Subsystem flag	17
	Dynamic bus control	18
	Terminal flag	19
	Parity	20

Command word: generated by a BC and directed at an RT to perform a specific action, which could be either to transmit or receive some data.

The remaining fields contain the address of the RT for which the word is destined (RT address/Terminal address), the direction of the data transmission (1 for transmit and 0 for receive), the word count or the mode value if applicable, and the sub-address or an indication of whether the command is a mode code. The sub-address is an integer between 0 and 31, which points to the data buffer where the data will be transmitted to or received from by the RT.

The RT address can contain up to 31 addresses in the range 00000B to 11110B. The last address slot 11111B is used only for broadcast command.

Mode codes are special commands used to alter the status or operation of the RT, such as request for self-test, shutdown, synchronization, etc.

Data word: contains 16 bits of data corresponding to the actual data being transmitted by the RT, and it does not follow a prescribed structure.

Status word: used by RTs to respond to commands issued by the BC and enables RT to communicate their status or error information to the BC after executing prescribed action (e.g., upon receiving a valid message) using various flags.

2.2.3 Communication Formats

The BC orchestrates the communications between RTs by following a strict, predefined order and timing. There are 4 types of communication instances which can take place between the components.

***BC-RT/RT-BC* communication**:

Receive (BC → RT):

1. BC sends a receive command to RT
2. BC sends data words
3. Designated RT sends a status word

Transmit (RT → BC)

1. BC sends a transmit command to RT
2. RT responds by sending a status word
3. RT sends data words

***RT-RT* communication**:

1. BC sends receive command to receiving RT
2. BC sends transmit command to sending RT
3. Sending RT sends a status word and follow up by sending the data words
4. Receiving RT then sends a status word

Mode code communication: The BC can send a mode code to a specific RT or several RTs by setting the sub-address/mode field to 00000B or 11111B (i.e., SA 0 or SA 31) and the word count field to the mode code value.

Broadcast communication: done only for messages in which the BC is the transmitting component, and all other components are on the receiving end.

The BC can send a broadcast message by setting the RT address field to 11111B, the broadcast address (i.e., TA 31) and eliminating the status words transmission for all the receiving RTs. The broadcast message will be sent to all remote terminals (that implement the broadcast option), but no remote terminals will respond to the message to avoid conflicts on the bus.[2] However, the lack of explicit response from the terminals means the message cannot be resent in case of errors.

2.2.4 Timing Constraints

Messages are transmitted over the bus by XORing the data with a 1 MHz clock and deriving from the signal the bit values. Bit values of 1 and 0 correspond to high to negative and negative to high voltage transitions, respectively.

[2] Most MIL-STD-1553 avoids using broadcast messages, as such they are not used on most of the MIL-STD-1553 systems, because these violate one of the key principles of 1553, which is that all successful data transfers be explicitly acknowledged before the next transfer is attempted. This is impossible to achieve in a broadcast environment.

Transmission of a command word from BC to RT results in the following response time constraints:

- Expected response time: $4\,\mu s \leq T \leq 12\,\mu s$
- If response time $12\,\mu s < T \leq 14\,\mu s$ then generate *Late Response* error.
- Else if $14\,\mu s < T$ then generate *No Response* error.

All data words that are part of the same message must be transmitted in sequence without any time gap between them. However, a time gap of 4 μs must be maintained between the end of a message and the start of another message; so an inter message time gap of 4 μs must be maintained by the BC.

Message transmission scheduling is structured around the major and minor frames. Messages are grouped in minor frames, which in their turn are grouped in major frames. The transmission frequency for minor frames is defined for each major frame. Aperiodic messages are inserted in the schedule according to their triggering events by appending them appropriately at the end of the next minor frames.

3 Mil-Std-1553 Attack Vectors

3.1 Assumptions and Attacker Position/foothold on 1553 Platform

The feasibility of a specific attack vector on the 1553 data bus depends on the type of access the attacker has on the platform. The main assumption made for most of the attacks and that we will consider as basis for our simulations is that the attacker can connect a rogue device on the network that is capable to statistically learn the frame schedule by exploiting the centralized multiplexing and scheduling scheme which is controlled by the BC.

There are 4 kinds of positions for the attacker:

1. Attacker controls or compromise an external system that uses data transmitted through the 1553
2. Attacker controls a RT that is connected to the 1553 network
3. Attacker controls a component that performs the function of bus controller on the bus
4. A combination of some of the above footholds.

Due to its centralized role, a compromised bus controller gives the attacker the capability to initiate new messages, and remove, modify, or replace existing messages.

A compromised remote terminal can initiate new messages, impersonate any of the existing genuine remote terminals as well as the bus controller itself, and disrupt message exchange between other terminals.

A compromised host can disrupt messaging between the terminals connected to the bus by violating deliberately the constraints and rules of the 1553 standard or by exchanging fake data/commands within/outside the regular bus schedule.

3.2 Attack Vectors and Types

Under the abovementioned assumption, different attack vectors are applicable in the 1553 network. Some of the attack vectors overlap or are dependent on other attack vectors [2, 3, 8].

The following is a non-exhaustive list of attack vectors which can be executed against the 1553 network:

1. Reconnaissance
    a. Identify the bus schedule
        i. Identify message timing and order
        ii. Identify message content
    b. Identify the bus topology and components
2. Injection
    a. Jamming
    b. Basic injection
    c. Selective injection
3. Impersonation/masquerade
4. Data leakage
5. Denial of service (DOS)
6. Frame overflow
7. Logic attacks

We describe below example attacks that fall under the abovementioned vectors.

Broadcast message spoofing: consists of leveraging the broadcast feature of 1553 to send fake broadcast messages to the terminals connected to the bus. This requires compromising the bus controller and having it send the broadcast messages to the terminals connected to the bus.

Mil-STD-1553 Network scanning: consists of probing the components connected to the bus by sending them benign messages and analyzing the response to infer or extract information about their status. This may be carried out by a compromised component impersonating the bus controller, or a compromised remote terminal interacting with other components without coordination with the bus controller.

Denial of service (*DOS*): results in blocking communication on the bus among the connected terminals. DOS may result from physical damage to the bus or some of the key connected components such as the BC, or by corrupting the output of some of the components or making them unresponsive when some response would be expected.

DOS can be executed through random injection on the bus of messages that will create some collisions. This can be achieved by transmitting words that correspond to some of the messages in accordance with the bus schedule. This will confuse the centralized multiplexing and scheduling, causing some decoding error, which may lead the BC to switch the transmission to a backup channel. In this case, for maximum effect, the attack could be carried out simultaneously against both the primary and backup channels. At the same time, to evade detection the attack must be conducted stealthily by minimizing the amount of confusion or noise injected.

An example DOS attack scenario is as follows:

1. Rogue device cause collision by sending periodically the same command word issued by the BC for specific RTs.
2. At the same, the device will respond on behalf of the RT, sending thereby fake information.
3. Collided messages will not be decoded by the RT or will fail message validation with no response from the RT. Such lack of response will enable the rogue device to fill the vacuum by responding on behalf of the RT with false information.

Injection: three kinds of injection attacks can be executed, including jamming, basic injection and selective injective, as described below:

- Jamming: consists of injecting many packets to slow down or block communication on the bus. Rather than targeting 100% utilization, it is advised to target a much lower utilization, e.g., 30–50%, to evade detection. This will slow down considerably the bus traffic while allowing basic functions such as power management to be available.
- Injection: consists of injecting packets selectively in the bus so as to minimize collisions with the scheduled traffic. This requires leveraging the periodicity in the bus schedule to predict the injection times that would achieve lower collision probability.
- Selective jamming: consists of injecting packets in the bus so as to collide repeatedly with specific message, thereby preventing two or more components from communicating over the bus.

Spoofing: consists of subverting or abusing the logic underlying the expected functioning of the bus system. For instance, this may consist of identifying empty slots in the bus schedule when no communication is expected to take place and using these slots to exchange malicious information. This takes advantage of the fact that there is no reliable way of determining the identity of the devices that are connected to the bus. So, a rogue device can masquerade as a legitimate component and confuse genuine terminals.

The attack may consist of modifying legitimate messages exchanged over the bus or sending fake messages within unexpected time slots (e.g., idle slots) or over unusual order.

Data leakage: consists of illegal data transfer between components, e.g., from a legitimate component connected to the bus with higher (security) classification level to an external component with lower classification. Data leakage can occur by changing the terminal address field, which allows directing the data to another component, by increasing the word count, which allows sending excess data, or by stuffing the leaked data through reserved bits in a status word.

Frame overflow: can occur unintentionally due to inadequate design of the bus schedule. However, it can also be induced intentionally to attack the 1553 network.

Here, the attack leverages the priority feature associated with aperiodic messages. Aperiodic messages may be executed with high or low priority.

Low priority messages are executed in the time frame between the last message in a minor frame and the start of the next minor frame.

High priority messages are executed immediately after completion of the current messages in the periodic bus list. The execution of periodic messages in the bus list resumes after completing these high priority aperiodic messages. However, unlike high priority messages, low priority messages are executed only if there is enough time to transmit these messages before the start of the next minor frame.

The attack consists of inducing minor frame overflow by generating high priority aperiodic messages that would delay the execution of periodic messages. The goal is to stretch the execution of the aperiodic messages in the bus list beyond the start of the next minor frame.

Low priority messages can also be targeted by padding the periodic messages list so that the time left in to process the low priority messages would not be enough to process these messages. This would block the low priority messages from ever running.

Logic attacks: consist of abusing or pushing to the extreme the constraints imposed by the 1553 standard on the bus and the connected components.

An example of such attack consists of creating a loop that exceeds the minor frame time defined in the major frame specification. Eventually, such a loop would lock up the 1553 bus such that the corresponding minor frame is never exit.

Another example consists of sending a broadcast message through a rogue RT.

(RT- > RT) with an unusual (in the context) mode code, e.g., sending a shutdown or busy mode code to all legitimate terminals during operation.

Normal RT- > RT messages are common message types, but RT- > RT Broadcast messages are relatively rare for most systems.

4 Simulation and IDS Dataset Generation

4.1 Simulation Setup

Our simulation setup involves the following main components:

- ABACO R15-USB-2M: MIL-STD-1553 multi-function, two dual-redundant channel, USB interface box, 8 Bi-directional discrete.
- Abaco BUSTOOLS/1553 GUI Software for MIL-STD-1553 bus analysis, simulation and data logging running on Microsoft Windows.

Abaco Systems R15-USB is a high-speed USB 2.0 interface with dual-redundant MIL-STD-1553A/B channels, which can operate either in dual-function or multi-function modes. In Dual-function mode, it operates simultaneously as either a Bus Monitor and Bus Controller, or a Bus Monitor with up to 31 Remote terminals. In Multi-function mode, it operates simultaneously as a Bus Controller, with up to 31 Remote Terminals and Bus Monitor. Our simulations were based on the multi-function mode.

BusTools/1553 provides a GUI to simulate completely a dual redundant MIL-STD-1553 data bus, including the Bus Controller, multiple remote terminals (up to 32), and bus monitor. In addition, BusTools/1553 provides analysis and debugging features including error injection and detection, data filtering, and bus data visualization.

4.2 Baseline Scenarios and Datasets

The ABACO BusTools simulation software provides several sample topologies which cover different avionics operational settings. We use one of these topologies to define our core baseline scenario for simulation and dataset generation. The baseline topology is depicted by Fig. 1. From the core bus list, we generate data samples representative of normal activities, and then by simulating different attack scenarios we generate attack samples.

The baseline setup consists of four remote terminals (RT1–RT4) and a bus controller. The four remote terminals consist of an inertial reference unit (IRU),

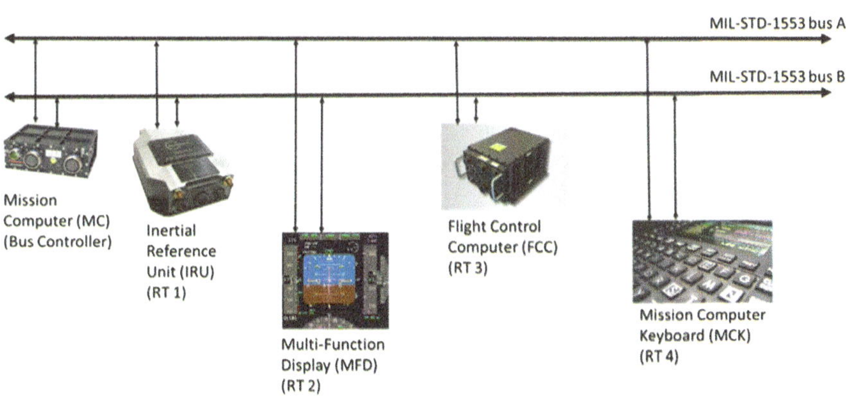

Fig. 1 Baseline architecture

a multi-function display unit (MFD), a flight control computer (FCC) and a mission computer keyboard (MCK).

- The bus controller is a Mission Computer (MC) that manages the data transfers among the different components.
- RT1 is an IRU which provides navigation data to the FD and FCC under the control of the MC.
- RT2 is a MFD unit which displays flight data from components such as MC, FCC or IRU.
- RT3 is an FCC, which produces flight status information and receives navigation and flight path data from the IRU and the MC, respectively.
- RT4 is a MCK that receives input from the crew and transmit such data to the MC.

Table 2 depicts the messages involved in the bus list; 18 messages are defined in total.

Table 3 depicts the data words defined within the messages. Table 4 depicts the bus list. The bus list consists of a major frame performed at a rate of 1 Hz, including 4 minor frames performed at a rate of 4 Hz.

Since, errors are to be expected in the regular operation of avionics system, one of the IRU navigation data buffers (i.e., in the simulation rounds) has an associated

Table 2 Message list

RT	Message description	SA	T or R	WC
IRU (T1)	Synchronize with data mode code	0	R	17
IRU (T1)	Navigation data output	1	T	9
IRU (T1)	Data wrap	20	both	1
IRU (T1)	Periodic built-in test (PBIT) results	30	T	32
MFD (RT2)	Synchronize with data mode code	0	R	17
MFD (RT2)	Nav data display input	2	R	9
MFD (RT2)	Data wrap	20	both	1
MFD (RT2)	Periodic built-in test (PBIT) results	30	T	20
FCC (RT3)	Synchronize with data mode code	0	R	17
FCC (RT3)	Navigation data input	1	R	9
FCC (RT3)	Display data selection input	2	R	2
FCC (RT3)	Data wrap	20	both	1
FCC (RT3)	Periodic built-in test (PBIT) results	30	T	32
MCK (RT4)	Synchronize with data mode code	0	R	17
MCK (RT4)	Data available	5	T	1
MCK (RT4)	Keyboard data	6	T	32
MCK (RT4)	Data wrap	20	both	1
MCK (RT4)	Periodic built-in test (PBIT) results	30	T	1

Table 3 Message data definitions

RT	Message description	T/R	SA	Wrd	Label	Units
IRU (RT1)	Synchronize with data mode code	R	0	1	Minor frame	Count
IRU (RT1)	Navigation data output	T	1	1–2	Altitude	Feet
IRU (RT1)	Navigation data output	T	1	3	Air speed	Knots
IRU (RT1)	Navigation data output	T	1	4	Heading	Degrees
IRU (RT1)	Navigation Data Output	T	1	5–6	Latitude	Degrees
IRU (RT1)	Navigation data output	T	1	7–8	Longitude	Degrees
IRU (RT1)	Navigation data output	T	1	9	Data valid	Boolean
IRU (RT1)	Data wrap	T	20	1	IRUWrapR	Counts
IRU (RT1)	Data wrap	R	20	1	IRUWrapT	Counts
IRU (RT1)	Periodic built-in test results	T	30	1	IRU bit OK	Boolean
MFD (RT2)	Synchronize with data mode code	R	0	17	Minor frame	Count
MFD (RT2)	Display input	R	2	1–2	Altitude	Feet
MFD (RT2)	Display input	R	2	3	Air speed	Knots
MFD (RT2)	Display input	R	2	4	Heading	Degrees
MFD (RT2)	Display input	R	2	5–6	Latitude	Degrees
MFD (RT2)	Display input	R	2	7–8	Longitude	Degrees
MFD (RT2)	Display input	R	2	9	Data valid	Boolean
MFD (RT2)	Data wrap	R	20	1	MFDWrapT	Counts
MFD (RT2)	Data wrap	T	20	1	MFDWrapR	Counts
MFD (RT2)	Periodic built-in test results	T	30	1	MFD bit OK	Boolean
FCC (RT3)	Synchronize with data mode code	R	0	17	Minor frame	Count
FCC (RT3)	Navigation data input	R	1	1–2	Altitude	Feet
FCC (RT3)	Navigation data input	R	1	3	Air speed	Knots
FCC (RT3)	Navigation data input	R	1	4	Heading	Degrees
FCC (RT3)	Navigation data input	R	1	5–6	Latitude	Degrees
FCC (RT3)	Navigation data input	R	1	7–8	Longitude	Degrees
FCC (RT3)	Navigation data input	R	1	9	Data valid	Boolean
FCC (RT3)	Display data selection input	R	2	1	Selection	1/0
FCC (RT3)	Data wrap	R	20	1	FCCWrapR	Counts
FCC (RT3)	Data wrap	T	20	1	FCCWrapT	Counts

(continued)

Table 3 (continued)

RT	Message description	T/R	SA	Wrd	Label	Units
FCC (RT3)	Periodic built-in test (PBIT) results	T	30	1	FCC bit Ok	Boolean
MCK (RT4)	Synchronize with data mode code	R	0	17	Minor frame	Count
MCK (RT4)	Data available	T	5	1	Data avail	Boolean
MCK (RT4)	Keyboard data	T	6	1	Key data	ASCII
MCK (RT4)	Data wrap	R	20	1	MCKWrapR	Counts
MCK (RT4)	Data wrap	T	20	1	MCKWrapT	Counts
MCK (RT4)	Periodic built-in test results	T	30	1	MCK bit OK	Boolean

error specified. An unintentional functional error is not a malicious occurrence, and therefore an anomaly detector should be able to distinguish unintentional errors from malicious errors.

To generate our baseline (normal) dataset, we run the simulation based on the abovementioned bus list for 10 min. Figures 2 and 3 show sample simulated data.

The simulation resulted in the generation of 23,000 legitimate message samples. As mentioned above, some of the generated (legitimate) messages have error flags associated with them.

The captured data was saved in.*bdmx* format, and an ASCII dump was generated as shown in Fig. 4.

Then, a parser was implemented to convert the ASCII dump in CSV format (see Fig. 5), which is suitable to conduct the data analysis and the design of the machine learning algorithms for anomaly detection.

The dataset consists of 55 fields described in Table 5.

4.3 Attack Scenarios and Datasets

To generate the attack samples, we executed 6 different attack vectors described in the following.

Attack 1: consists of a DOS attack where a rogue terminal (RTO) targets the FCC (RT3) by sending random words in a loop. The attack was run for 30 s and generated 148 messages, a mix of normal and attack data. Sample messages are shown in Fig. 6.

Attack 2: consists of a DOS were broadcast messages were sent in a loop by the rogue terminal (RTO). By running the attack for 30 s., 373 messages were generated, involving both legitimate and malicious data. Sample messages are shown in Fig. 7.

Attack 3: consists of fake data injection. A rogue terminal (RTO) replicates one of the messages (i.e., Navigation Data Output) sent by one of the legitimate RTs (RT1–IRU)

Table 4 Bus list

Minor frame	Message type	Source RT/SA	Destination RT/SA	WC	Description	Message name
1,2,3,4	MC 17	BC	1/0	17	Sync with data	Mode code 17—1—x
1,2,3,4	MC 17	BC	2/0	17	Sync with data	Mode code 17—2—x
1,2,3,4	MC 17	BC	3/0	17	Sync with data	Mode code 17—3—x
1,2,3,4	MC 17	BC	4/0	17	Sync with data	Mode Code 17—4—x
1	BC - > RT	BC	1/20	1	Data wrap	WrapIRU-20-R
2	BC - > RT	BC	2/20	1	Data wrap	WrapMFD-20-R
1	BC - > RT	BC	3/20	1	Data wrap	WrapFCC-20-R
2	BC - > RT	BC	4/20	1	Data wrap	WrapMCK-20-R
3	RT - > BC	1/20	BC	1	Data wrap	WrapIRU-20-T(B)
4	RT - > BC	2/20	BC	1	Data wrap	WrapMFD-20-T(B)
3	RT - > BC	3/20	BC	1	Data Wrap	WrapFCC-20-T(B)
4	RT - > BC	4/20	BC	1	Data wrap	WrapMCK-20-T(B)
1	RT - > BC	1/30	BC	32	PBIT results	PBTIRU-30-T
2	RT - > BC	2/30	BC	20	PBIT results	PBTMFD-30-T
3	RT - > BC	3/30	BC	32	PBIT results	PBTFCC-30-T
4	RT - > BC	4/30	BC	1	PBIT results	PBTMCK-30-T
1,2,3,4	RT - > BC	4/5	BC	1	Data available	DAvailMCK-5-T
1,2,3,4	RT - > BC	4/6	BC	32	Keyboard data	DataMCK-6-T-Cond
1	BC - > RT	BC	3/2	2	Data selection	Sel-FCC-2-R
4	RT - > BC	3/2	BC	10	FCC data out	Sel-FCC-data-2-T
1,2,3,4	RT - > RT	1/1	3/1	9	Nav data	Nav-data-to-FCC
1,2,3,4	RT - > RT	1/1	2/2	9	Nav data	Nav-data-to-MFD

by slightly altering one of the data items (Altitude). By running the attack for 30 s, 970 messages are generated (malicious and normal). Sample messages are shown in Fig. 8.

Attack 4: consists of fake data injection like attack 3, except that in this case fake data is generated. This is less subtle and noisier, but its impact can be felt much quicker. By running the attack for 30 s, 970 messages are generated (malicious and normal). Sample messages are shown in Fig. 9.

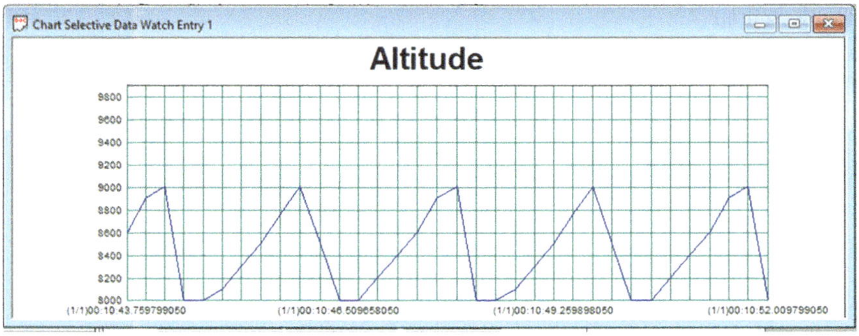

Fig. 2 Sample simulated data during normal operations

Fig. 3 Sample simulated messages during normal operations

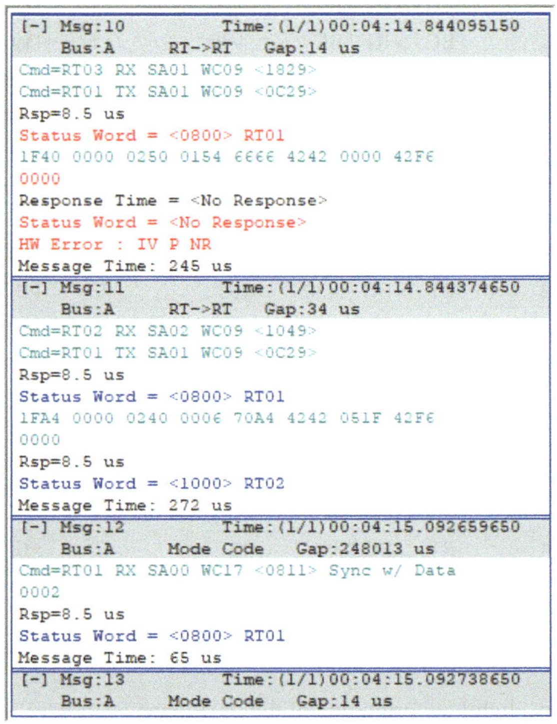

Attack 5: logic attack consisting of unusual broadcasting. In this case a rogue RT broadcast mode code value 4 (0×04), which corresponds to Transmitter Shutdown. It is unusual to issue such a request in the middle of an operation (i.e., during flight). The attack ran for 30 s and generated 981 messages, a mix of normal and malicious samples.

```
[-] Msg:2          Time:(1/1)00:04:14.842738150
      Bus:A        Mode Code    Gap:14 us
Cmd=RT02 RX SA00 WC17 <1011> Sync w/ Data
0001
Rsp=8.5 us
Status Word = <1000> RT02
Message Time: 65 us
```

(a) Example data from the .bmdx capture for Message #2

```
Msg:2           Time Tag:(1/1)00:04:14.842738150
Bus:A           Mode Code    Gap:14 us
Cmd=02-RX-00-17 <1011> Sync w/ Data
0001
Rsp=8.5 us
Sts=<1000>
Message Time: 65 us
```

(b) Message #2 in the ASCII dump

Fig. 4 Sample message viewed in BusTools GUI and corresponding ASCII dump

	A	B	C	D	E	F	G	H	I	J	K	L	M	N
1	msgId	timestam	error	modeCod	channel	connType	sa	ssa	da	dsa	wc	modeCod	txRsp	txSts
2	1	254.8427	FALSE	TRUE	A	BC->RT	N/A	N/A	1	0	N/A	17	N/A	N/A
3	2	254.8427	FALSE	TRUE	A	BC->RT	N/A	N/A	2	0	N/A	17	N/A	N/A
4	3	254.8428	FALSE	TRUE	A	BC->RT	N/A	N/A	3	0	N/A	17	N/A	N/A
5	4	254.8429	FALSE	TRUE	A	BC->RT	N/A	N/A	4	0	N/A	17	N/A	N/A
6	5	254.8431	FALSE	FALSE	A	BC->RT	N/A	N/A	1	20	1	N/A	N/A	N/A
7	6	254.8431	FALSE	FALSE	A	BC->RT	N/A	N/A	3	20	1	N/A	N/A	N/A
8	7	254.8432	FALSE	FALSE	A	RT->BC	4	5	N/A	N/A	1	N/A	8.5	0x2000
9	8	254.8433	FALSE	FALSE	A	RT->BC	1	30	N/A	N/A	32	N/A	8.5	0x0800

O	P	Q	R	S	T	U	V	W	X	Y	Z	AA
rxRsp	rxSts	dw0	dw1	dw2	dw3	dw4	dw5	dw6	dw7	dw8	dw9	dw10
8.5	0x0800	0x0001	N/A	N/A	N/A	N/A	N/A	N/A	N/A	N/A	N/A	N/A
8.5	0x1000	0x0001	N/A	N/A	N/A	N/A	N/A	N/A	N/A	N/A	N/A	N/A
8.5	0x1800	0x0001	N/A	N/A	N/A	N/A	N/A	N/A	N/A	N/A	N/A	N/A
8.5	0x2000	0x0001	N/A	N/A	N/A	N/A	N/A	N/A	N/A	N/A	N/A	N/A
8.5	0x0800	0x0000	N/A	N/A	N/A	N/A	N/A	N/A	N/A	N/A	N/A	N/A
8.5	0x1800	0x0000	N/A	N/A	N/A	N/A	N/A	N/A	N/A	N/A	N/A	N/A
N/A	N/A	0x0000	N/A	N/A	N/A	N/A	N/A	N/A	N/A	N/A	N/A	N/A

AB	AC	AD	AE	AF	AG	AH	AI	AJ	AK	AL	AM	AN	AO
dw11	dw12	dw13	dw14	dw15	dw16	dw17	dw18	dw19	dw20	dw21	dw22	dw23	dw24
N/A	N/A	N/A	N/A	N/A	N/A	N/A	N/A	N/A	N/A	N/A	N/A	N/A	N/A
N/A	N/A	N/A	N/A	N/A	N/A	N/A	N/A	N/A	N/A	N/A	N/A	N/A	N/A
N/A	N/A	N/A	N/A	N/A	N/A	N/A	N/A	N/A	N/A	N/A	N/A	N/A	N/A
N/A	N/A	N/A	N/A	N/A	N/A	N/A	N/A	N/A	N/A	N/A	N/A	N/A	N/A
N/A	N/A	N/A	N/A	N/A	N/A	N/A	N/A	N/A	N/A	N/A	N/A	N/A	N/A
N/A	N/A	N/A	N/A	N/A	N/A	N/A	N/A	N/A	N/A	N/A	N/A	N/A	N/A
N/A	N/A	N/A	N/A	N/A	N/A	N/A	N/A	N/A	N/A	N/A	N/A	N/A	N/A

AO	AP	AQ	AR	AS	AT	AU	AV	AW	AX	AY	AZ
dw24	dw25	dw26	dw27	dw28	dw29	dw30	dw31	malicious	injected	gap	msgTime
N/A	N/A	N/A	N/A	N/A	N/A	N/A	N/A	FALSE	FALSE	N/A	65
N/A	N/A	N/A	N/A	N/A	N/A	N/A	N/A	FALSE	FALSE	14	65
N/A	N/A	N/A	N/A	N/A	N/A	N/A	N/A	FALSE	FALSE	14	65
N/A	N/A	N/A	N/A	N/A	N/A	N/A	N/A	FALSE	FALSE	14	65
N/A	N/A	N/A	N/A	N/A	N/A	N/A	N/A	FALSE	FALSE	99	65
N/A	N/A	N/A	N/A	N/A	N/A	N/A	N/A	FALSE	FALSE	14	65
N/A	N/A	N/A	N/A	N/A	N/A	N/A	N/A	FALSE	FALSE	14	65

Fig. 5 Sample data from normal dataset in CSV

Table 5 Dataset format

Fields	Description
msgID	Sequence number associated with the message by the simulator
timestamp	Message timestamp in seconds
Error	Indicate whether or not the error bit of the status word related to the message is set: contains TRUE/FALSE accordingly
modeCode	Indicate whether or not the message is a mode command message: contains TRUE/FALSE accordingly
Channel	Channel associated with the message
connType	Communication type
sa	Address of sending RT
ssa	Sub-address of the sending subsystem from the sending RT
da	Address of receiving RT
dsa	Sub-address of the receiving subsystem at the receiving RT
wc	Word count: number of data words included in the message
modeCode value	Mode code value when applicable
txRsp	Transmit command response time in μs
txSts	Transmit status word
rxRsp	Receive command response time in μs
rxSts	Receive status word
dw0 … dw31	Values of the data word included in the message ranging from dw0 to dw31; N/A is used when there is no data words for a field
Malicious	Indication of whether or not the message is malicious: contains TRUE/FALSE accordingly
Injected	Indication of whether or not part of the data included in the message is injected: contains TRUE/FALSE accordingly
Gap	Inter-message gap time in μs
msgTime	Message time in μs

Attack 6: a combination of attacks 4 and 5. The attack is run for 30 s and 1004 messages are generated consisting of a mix of malicious and normal messages.

5 Conclusion

Data is one of the most essential ingredients in the design of an IDS, especially when the corresponding model depends on machine learning techniques.

By setting up a simulation environment, we have identified and executed different attack scenarios. Using the same setup, more scenarios can be executed. But we believe the example scenarios generated so far provides a good basis toward designing and evaluating IDS for MIL-STD-1553 platforms. The generated dataset is available

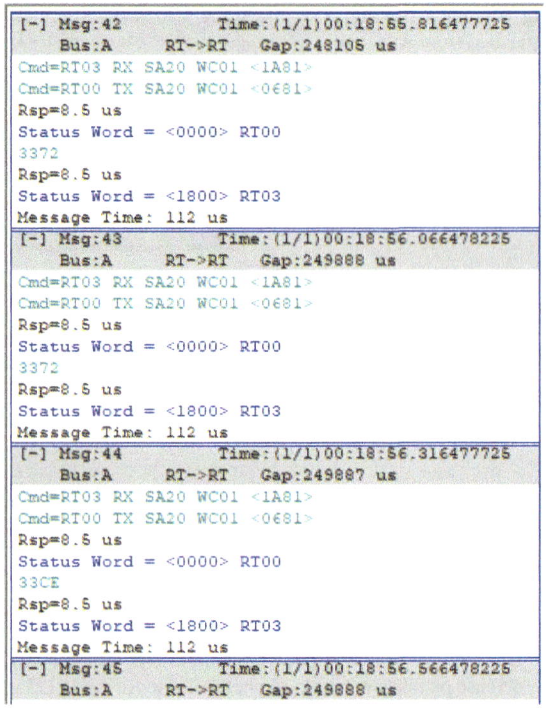

Fig. 6 Sample messages generated from Attack 1

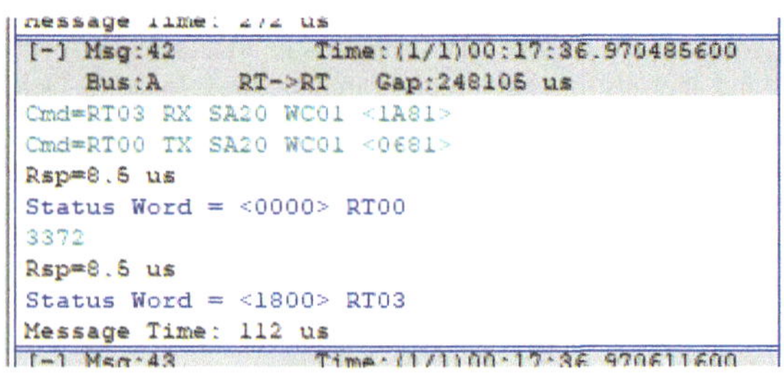

Fig. 7 Sample messages generated from Attack 2

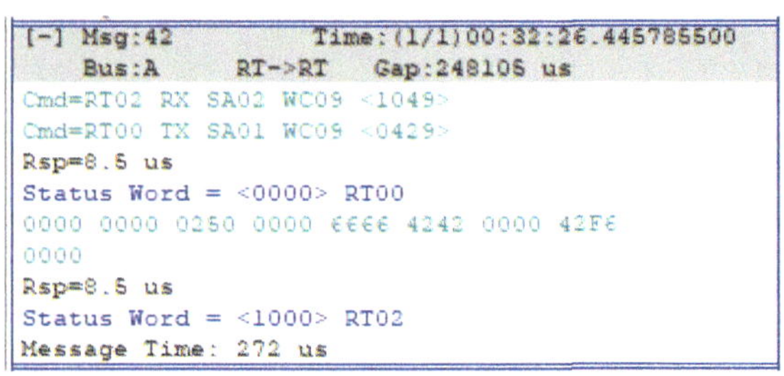

Fig. 8 Sample messages generated from Attack 3

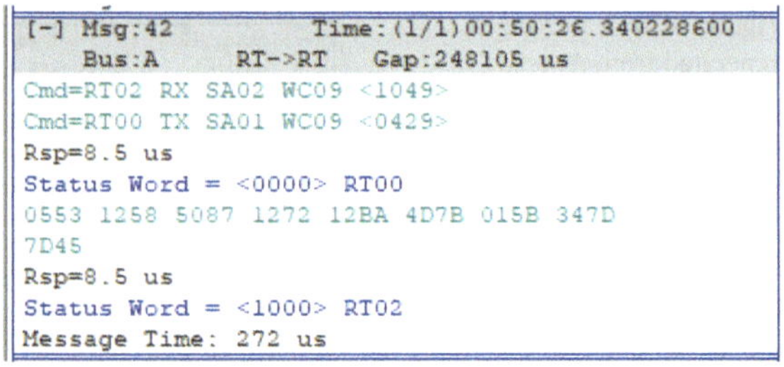

Fig. 9 Sample messages generated from Attack 4

for public use through the website of the Information Security and Object Technology (ISOT) Lab at https://www.uvic.ca/ecs/ece/isot/datasets/index.php

References

1. EDITION, S. Mil-std-1553 designer's guide
2. Nguyen TD (2015) Towards mil-std-1553b covert channel analysis. Tech. rep., Monterey, California. Naval Postgraduate School
3. Hayden PM, Woolrich DK, Sobolewski KD (2017) Providing cyber situational awareness on defense platform networks. Cyber Defense Rev 2(2):125–140
4. Losier B (2019) Design of a time-based intrusion detection algorithm for the MIL-STD-1553. Master thesis, Royal Military College of Canada, Kingston, Ontario, Canada, January 2019
5. Mcgraw RM, Fowler MJ, Umphress D, Macdonald RA (2014) Cyber threat impact assessment and analysis for space vehicle architectures. In: SPIE Defense+ Security (2014), International Society for Optics and Photonics, pp. 90850K–90850K
6. Stan O, Elovici Y, Shabtai A, Shugol G, Tikochinski R, Kur S (2017) Protecting military avionics platforms from attacks on MIL-STD-1553 Communication Bus. arXiv:1707.05032v1[cs.CR] 17 Jul 2017
7. Yahalom R, Barishev D, Steren A, Nameri Y, Roytman M, Porgador A, Elovici Y (2019) "RT spoofing attacks on MIL-STD-1553 communication traffic", Mendeley Data v3. Elsevier. https://doi.org/10.17632/jvgdrmjvs3.3
8. Huitsing P, Chandia R, Papa M, Shenoi S (2008) Attack taxonomies for the Modbus protocols. Int J Crit Infrastruct Prot 1:37–44. https://doi.org/10.1016/j.ijcip.2008.08.003

Unsupervised Anomaly Detection for MIL-STD-1553 Avionic Platforms Using CUSUM

Krunal Sachdev, Hadeer S. Saad, Issa Traore, Karim Ganame, and Oussama Boudar

Abstract MIL-STD-1553 is a military standard developed by the US department of defense for communication among military avionic platforms (e.g., F-35 and F-16). It has been widely accepted worldwide for more than five decades and is used in many applications other than military avionics. It follows a strict and deterministic procedure for communication among its components. However, research has suggested that it has many vulnerabilities associated with it that can be exploited to carry a range of attacks on it. And since numerous applications make use of this standard, it is crucial to protect MIL-STD-1553 networks. This chapter presents an unsupervised anomaly detection scheme using the CUSUM algorithm for the MIL-STD-1553 protocol. A dataset was collected in the ISOT lab by executing six attack vectors on a simulated MIL-STD-1553 network. We leverage the time-based properties of the communication bus to extract a set of relevant features that are fed to the CUSUM algorithm for detection. The experimental evaluation of the proposed detector using the dataset yielded promising results, which are very encouraging considering the unsupervised nature of the underlying algorithm.

Keywords Unsupervised anomaly detection · MIL-STD-1553 · Protocol · Avionic platform · CUSUM algorithm · Dataset · Intrusion detection · Cybersecurity · Vulnerabilities · Simulation · Attack type · Feature correlation · Detection model

K. Sachdev · H. S. Saad · I. Traore (✉)
Department of Electrical and Computer Engineering, University of Victoria, Victoria, BC, Canada
e-mail: itraore@ece.uvic.ca

K. Sachdev
e-mail: krunalsachdev8495@gmail.com

H. S. Saad
e-mail: hadeer.sma@gmail.com

K. Ganame · O. Boudar
Efficient Protections Inc, Montreal, QC, Canada
e-mail: ganame@streamscan.io

O. Boudar
e-mail: oussama.boudar@streamscan.io

I. Traore et al. (eds.), *Artificial Intelligence for Cyber-Physical Systems Hardening*, Engineering Cyber-Physical Systems and Critical Infrastructures 2,
https://doi.org/10.1007/978-3-031-16237-4_5

1 Introduction

MIL-STD-1553 is a military standard introduced by the US Department of Defense (DoD) in 1973 [1]. It has also been widely used in other branches of armed forces other than military avionics and has served the military for more than 50 years by now. It is based on a master/slave mechanism where the master sends messages in a fixed and predefined time and order.

While it was designed with built-in reliability and fault tolerance capabilities, it has no inherent cybersecurity protections. The MIL-STD-1553 protocol was introduced at a time when cybersecurity was an afterthought. As a result, existing data buses are highly vulnerable. Furthermore, as more types of avionics data buses are being deployed and interconnected in both new and updated aircraft there is an increasing concern that these vulnerabilities in security could allow unauthorized access to devices communicating on these buses.

As highlighted in a master's thesis conducted by Blaine Losier at the Royal Military College (RMC) of Canada, classified research, recently performed, has exposed several vulnerabilities in the MIL-STD-1553 protocol [2]. For instance, in [3], the authors demonstrated the feasibility of cyber-attacks against MIL-STD-1553 buses used for communications between subsystems in a satellite.

While reports on cyber incidents targeted at defense platforms based on MIL-STD-1553 currently remain classified, published examples against similar protocols in other domains provide strong indicators of the feasibility of such attacks. Likewise, several vulnerabilities have been reported in the MODBUS serial protocol, which underlies existing supervisory control and data acquisition (SCADA) and industrial control systems (ICS) [4]. As serial buses, MODBUS and MIL-STD-1553 share many of the same characteristics, and as demonstrated in [5] many of the MODBUS vulnerabilities reported in [4] are applicable against MIL-STD-1553.

So, considering the use of MIL-STD-1553 across multiple avionics platforms, it is imperative to continue research to build an intrusion detection system (IDS) solution for the standard. This chapter presents a step toward developing an intrusion detection scheme based on unsupervised anomaly detection tailored for the MIL-STD-1553 protocol, which targets a broad range of attacks aimed at this platform.

The basic principle of intrusion detection assumes that intrusive activities are noticeably different from normal ones and thus are detectable. Many intrusion detection approaches have been proposed for conventional systems in the literature and in industry since the seminal work of Anderson in the 1980s [6]. Traditionally, intrusion detection techniques are broadly classified into two categories: misuse detection and anomaly detection. Misuse detection assumes that most attacks leave a set of signatures in the stream of communication and activity traces, and thus attacks are detectable if these signatures can be identified and matched against malicious system behaviors. However, misuse detection approaches are strictly limited to known attacks. They are ineffective when faced with new attack vectors or variants of known attacks.

Anomaly detection is based on the premise that all intrusive activities are anomalous [6–8]. Consequently, it involves establishing normal activity profiles for the system, and then tracking and reporting any deviation from normal profiles as an intrusion. The primary advantage of anomaly detection is its ability to detect novel attacks. However, if the underlying normalcy model is not carefully designed, the false alarm rate can be quite high.

Anomaly detection is suited for monitoring the MIL-STD-1553 bus due to its great level of determinism and predictability, which provide a good foundation for designing a strong model of the normal behavior of the system and tracking violations of such a model.

Existing proposals on detecting intrusions against the MIL-STD-1533 data buses are largely based on supervised machine learning models and cover only a limited set of attacks, such as covert channel or spoofing attacks [2, 9–11]. Our proposed approach is based on unsupervised machine learning techniques [12–14]. The detector monitors and analyzes bus traffic data and extracts a variety of features. The features are fed to the CUSUM algorithm to detect anomalous behavior. Evaluation is done using datasets collected at the Information Security and Object Technology (ISOT) Lab in a simulated environment.

The remainder of the chapter is structured as follows. Section 2 gives a brief overview of the datasets. Section 3 presents the extracted machine learning (ML) features that form the basis of the classification models. Section 4 present the proposed detection techniques and give a brief overview of the underlying algorithms. Section 5 presents the experimental evaluation of the proposed detection models and discuss the obtained results. Section 6 makes some concluding remarks and discusses future perspectives.

2 Datasets

The availability of adequate datasets is crucial to guide the design and evaluation of anomaly detection models that use machine learning techniques. We used datasets collected at the ISOT lab in a simulated environment consisting of normal activities and a series of attacks against the MIL-STD-1553 databus. We provide a brief overview of the datasets in the following.

The simulation was conducted using Abaco R15-USB-2 M USB interface box and Abaco BUSTOOLS/1553 GUI Software for MIL-STD-1553 bus analysis, simulation, and data logging.

Figure 1 depicts the simulated bus components. The example bus architecture used in the simulation consists of a bus monitor (BM) and five different avionic systems, including a flight control computer (FCC) as RT3, a mission computer (MC) as BC, an inertial reference unit (IRU) as RT1, a mission control keyboard (MCK) as RT4, and a multi-function display unit (MFD) as RT2. The MC serves as bus controller (BC), while the remaining components are remote terminals.

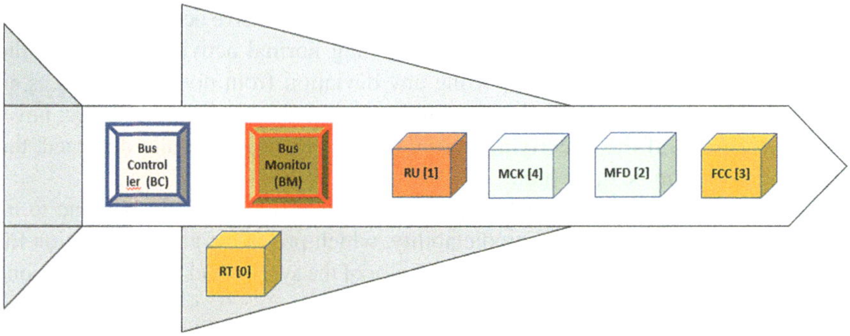

Fig. 1 Simulated MIL-STD-1553 bus components

The above components were the only ones used to execute the normal activities. To simulate the attack, we added a rogue terminal as RT0.

Our investigations uncovered several possible attack vectors against MIL-STD-1553; more details are provided about these attack vectors in the progress report. We selected a subset of these attacks to generate our datasets. A baseline dataset consisting of normal activities was generated by running the simulation for 10 min. Subsequently, 6 different attack scenarios were run separately against the baseline architecture, by introducing RT0 as rogue terminal. Table 1 provides a summary of the different datasets and Table 2 shows the different fields involved in the raw data.

As shown in the Table 1, the datasets obtained from executing the attacks include a mix of normal and attack messages. All the messages in the datasets were assigned adequate labels, clearly identifying them as legitimate or malicious. The raw datasets were parsed and converted into CSV format. Each dataset consists of 55 different fields. More details on the datasets can be found in Chap. 4.

Figure 2 shows the spread of the different type of samples in the data. The graph reveals an imbalance in the dataset between attack and benign samples. This is expected for security datasets in general, and IDS datasets specifically. By comparing different techniques to handle imbalanced data and running selected classical supervised machine learning algorithms (i.e., Random Forests, Support Vector Machines, Decision Tree, Extra Decision Tree, and Ada Boost) on the data, it was found that the imbalance does not have any impact on classification accuracy.

3 Features Model

To identify a set of useful features, we analyzed the statistical characteristics of the raw features depicted in Table 2. To analyze the correlation between different features a heatmap is created based on the correlation matrix of the dataset as shown in Fig. 3.

Table 1 Outline of the collected MIL-STD-1553 datasets

Dataset #	Type of data	Attack type	Attack description	Number of messages	Simulation length
1	Normal	N/A	N/A	23,000	10 min
2	Mix normal/malicious	Targeted/Basic DOS (Attack 1)	rogue terminal (RT0) targets the FCC (RT3) by sending random words in a loop	148	30 s
3	Mix normal/malicious	Multi-target DOS (Attack 2)	Rogue terminal (RT0) sends broadcast messages in a loop	373	30 s
4	Mix normal/malicious	Subtle Fake data injection (Attack 3)	rogue terminal (RT0) replicates one of the messages sent by one of the legitimate RTs by slightly altering one of the data items	970	30 s
5	Mix normal/malicious	Noisy fake data injection (Attack 4)	like the above attack, except that in this case fake data is randomly generated	970	30 s
6	Mix normal/malicious	Logic attack (Attack 5)	a rogue RT broadcast mode code value 4 (0 $\times$ 04), which corresponds to Transmitter Shutdown, and which is unusual	981	30 s
7	Mix normal/malicious	Hybrid Logic/fake data injection (Attack 6)	combination of the attacks in datasets 5 and 6	1004	30 s

Table 2 Raw dataset format

Fields	Description
msgID	Sequence number associated with the message by the simulator
timestamp	Message timestamp in seconds
error	Indicate whether or not the error bit of the status word related to the message is set: contains TRUE/FALSE accordingly
modeCode	Indicate whether or not the message is a mode command message: contains TRUE/FALSE accordingly
channel	Channel associated with the message
connType	Communication type
sa	Address of sending RT
ssa	Sub-address of the sending subsystem from the sending RT
da	Address of receiving RT
dsa	Sub-address of the receiving subsystem at the receiving RT
wc	Word count: number of data words included in the message
modeCode value	Mode code value when applicable
txRsp	Transmit command response time in μs
txSts	Transmit status word
rxRsp	Receive command response time in μs
rxSts	Receive status word
dw0 … dw31	Values of the data word included in the message ranging from dw0 to dw31; N/A is used when there is no data words for a field
Malicious	Indication of whether the message is malicious: contains TRUE/FALSE accordingly
injected	Indication of whether part of the data included in the message is injected: contains TRUE/FALSE accordingly
gap	Inter-message gap time in μs
msgTime	Message time in μs

Feature correlation analysis allows understanding the relationships between multiple attributes and features in a dataset. We can get some insight, such as whether any features depend on or cause another feature.

The heatmap provides a graphical representation of the data that uses a system of color-codes to represent different values. In the heatmap, correlation ranges from − 1 to + 1. Values closer to zero means there is no linear trend between the two variables. The close to 1 the correlation is the more positively correlated they are. That is as one increases so does the other and the closer to 1 the stronger this relationship is.

As it can be seen from the heatmap, there are strong correlations for several feature-pairs such as connType (communication type), modeCodeVal (mode code value), txRsp (transmit command response time), txSts (transmit status word), and rxRsp (receive command response time). We can also notice a relationship between words counts (WC) and [dw0 to dw31] (values of the data word).

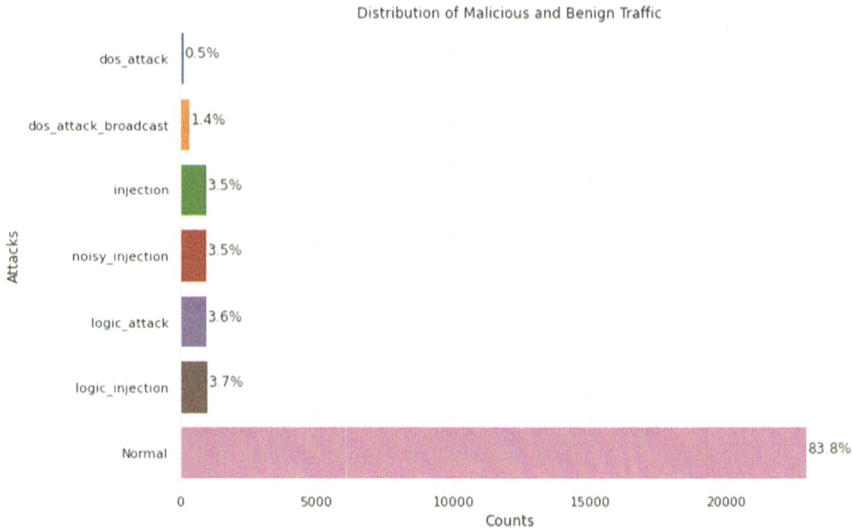

Fig. 2 Attack and benign sample distribution among the data

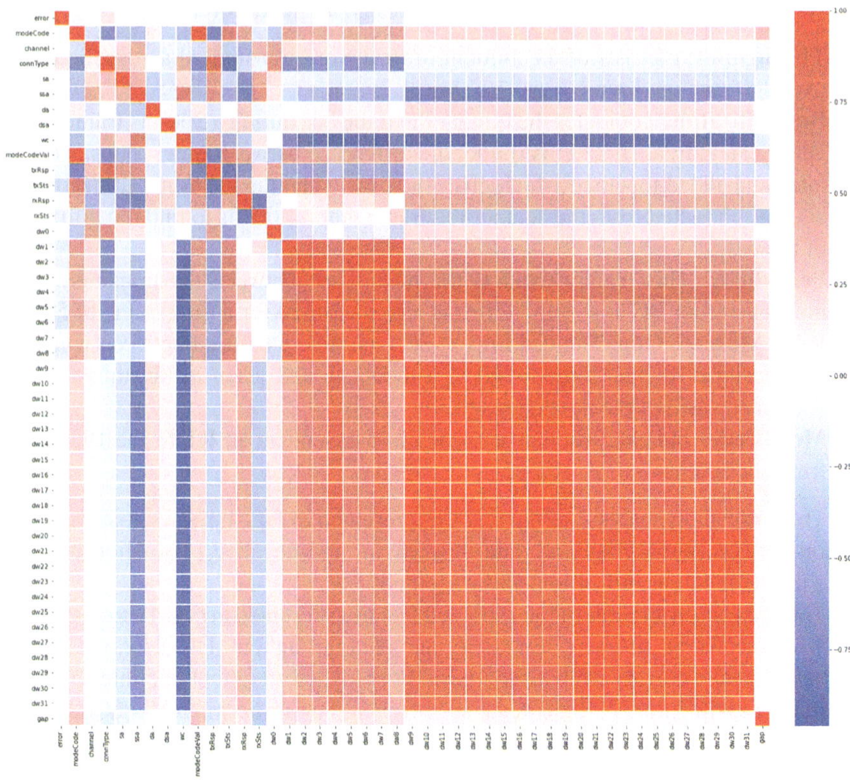

Fig. 3 Heatmap based on the correlation matrix of the dataset

Table 3 List of features extracted from the raw data

Features	Description	Values
Inter-message gap	Time interval between the end of a message (i.e., when the BC receives the corresponding final status word) and the start of the next message (i.e., when the corresponding first command is sent)	Numeric
Data throughput	Amount of data words transmitted over the bus during a specific time window δt; by default, we set $\delta t = 50$ ms	Ratio
Bus utilization	Percentage of time during which the bus is busy transmitting data measured over a specific time window δt; by default, we set $\delta t = 50$ ms	Percentage
Periodicity	The mean time between the BC sending messages to a specific RT, measured over a 50 ms frame	Numeric
modeCode	Indicate whether the message is a mode command message	True, False

From the data analysis, we identified a suite of five promising features depicted in Table 3. Two of the features are raw features and the other three features are obtained from transformation or combination of some of the raw features.

A brief description of the extracted features is provided as follows:

- **Throughput**: consists of the amount of data word transmitted across the bus in over 200 ms time frame.
- **Bus Utilization**: computes the utilization of the bus during a 200-ms frame. After getting an individual 200 ms frame, the difference between timestamps is calculated and added together, which gives the time when the bus is not utilized. This is then subtracted from the actual time frame that is 200 ms to find the utilization of the bus.
- **Periodicity**: This feature extracts the mean time between the transmission of messages from BC to specific RT. It is also measured for a time frame of 50 ms.
- **Inter Message Gap**: This feature extracts the time measured between when the BC receives the final status word of a message, and when it transmits the first command word of the next message.
- **ModeCode**: Indicate whether a mode code is sent by an RT to the BC over a frame of 200 ms.

The correlation analysis, of the features presented in Fig. 4, shows positive correlation between periodicity and data throughput, and negative correlation between data throughput, bus utilization, and periodicity.

The correlation analysis of the extracted features presented in Fig. 4 shows a strong positive correlation between periodicity and data throughput. However, we can see a slight positive correlation between periodicity/data throughput and inter message gap and bus utilization. A slight negative correlation exists between ModeCode value and other features such as throughput, inter message gap and periodicity. We should always be wary of the existence of perfect correlation. Suppose a dataset has perfectly positive or negative correlation. In that case, there is a high chance that the model's

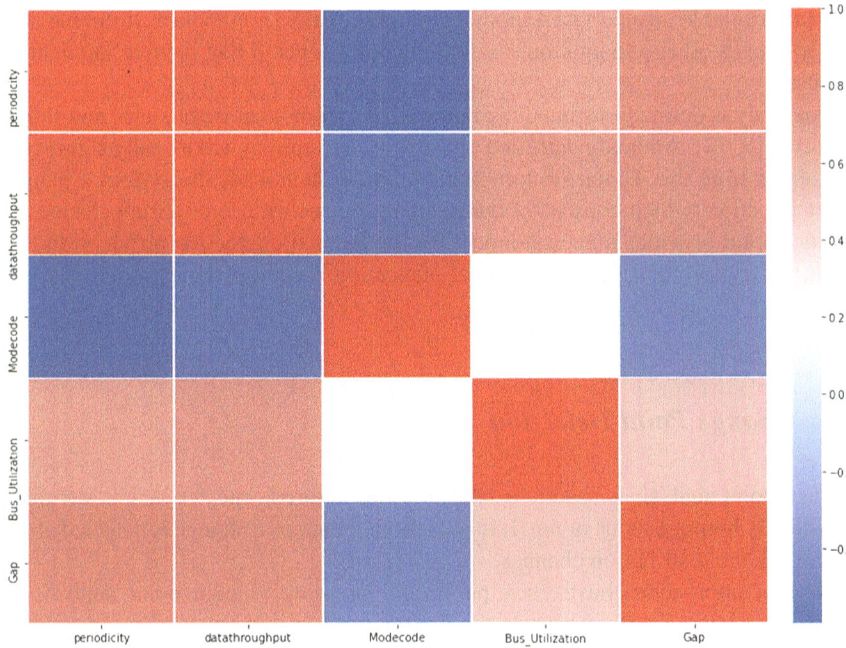

Fig. 4 Correlation among features

performance will be impacted by multicollinearity. This happens when one feature in a model can be linearly predicted from the others with a high degree of accuracy, leading to misleading results. However, this can be solved using algorithms such as decision trees and boosted trees algorithms which are immune to multicollinearity. Another solution is to remove some of the features to prevent this from happening.

4 Detection Model

Anomaly detection approaches fall into two subcategories: supervised and unsupervised. Supervised anomaly detection consists of creating a baseline for known benign network and system activity, and then looking for any events that seem anomalous with respect to the baseline. Supervised approaches require prior knowledge of normal instances which can be used to build the baseline.

Unsupervised approaches consist of identifying anomalous behaviors without any prior knowledge. In other words, while supervised approaches require some training on normal instances, unsupervised approaches do not require any training; detection can take place early on, without going through the process of building a baseline.

Most available unsupervised approaches use statistical methods to cluster the data closer to each other and look out for outliers/rare events to flag them as anomalous events.

Anomaly is defined as something that deviates from what is standard, normal, or expected [7, 13]. Anomaly detection consists of determining statistically significant deviations from the regular or normal pattern of behavior for the system. Change-point detection, which consists of determining the occurrence of abrupt changes in a data stream provides a natural model for the anomaly detection problem. In this work, we explore unsupervised anomaly detection based on change point analysis techniques.

4.1 Change Point Detection

Change point analysis is used to establish if a monitored time series data stream is statistically homogeneous or not. If it is not homogeneous a change is detected at the time when the distribution changes.

Several approaches have been published on using change point analysis to detect intrusions [17–22], however, these proposals have targeted only conventional networks.

To detect a change point, the data should be received sequentially or in a stream fashion, known as a sequence of observations. The detection occurs when a single or multiple changes suddenly happen in the distribution at the unknown change points T1, T2..., Tn, and between every pair of change points, it is usually assumed that the observations are independent and identically distributed.

There are two statistics that are commonly observed to detect a change: the mean and variance. The mean was used in the first change point detection model introduced by [23] for detecting a change in a univariate mean. Then, this approach was extended to detect a change in different setting, for example, in the variance [24], in multivariate change detection [25] and in non-parametric setting [26, 27] either for a single or multiple change points.

There are two main types of change point detection algorithms, namely, batch and sequential, also known as offline and online, respectively. In batch detection models the decision of homogeneity or a change point is made offline only after the entire set of observations are received. A fixed length sequence of observations is examined to determine whether any change point occurred at a particular point in the sequence or not. In contrast, in the sequential detection models the decision of homogeneity or a change point is made online while the observations are being received sequentially.

The offline method performs well with a sequence of observations that has only a small number of change points [28, 29]. This may not be adequate for MIL-STD-1553 environments where many devices are constantly generating data. Online methods are more adequate to handle such potentially unending stream of data. Therefore, in this work, we have decided to use online change detection approaches by exploring one of the most popular algorithms for online change detection called CUSUM [15, 16].

4.2 Using CUSUM Algorithm

As the name suggests, CUSUM stands for Cumulative Sum. It is statistical sequential change detection algorithm that was discovered by Page at the University of Cambridge in 1954 [30]. It is one of the most well-known algorithms for change detection. Change detection refers to finding changes in sequential data when a property of the time series changes [31]. The first classic CUSUM algorithm was designed for independent and identical distributions.

There are different forms of CUSUM as defined by Page [30]: direct or recursive forms and one-sided or two-sided forms. The aim of this algorithm, like any other change detection algorithm, is to detect the change when the state of the process changes from normal behavior to an abnormal one at a given time. In other words, a threshold is set for normal behavior, and if that threshold is exceeded, the algorithm sends an alarm at that time.

The algorithm to identify attacks based on CUSUM is written in Python programming language and uses one of the known Python modules for change detection in it.

Algorithm 1: CUSUM algorithm [32]

$$\begin{cases} s[t] = x[t] - x[t-1] \\ g^{+}[t] = max\big(g^{+}[t-1] + s[t] - drift, 0\big) \\ g^{-}[t] = max\big(g^{-}[t-1] + s[t] - drift, 0\big) \end{cases}$$

$$if g^{+}[t] > threshold OR g^{-}[t] > threshold$$

$$\begin{cases} t_{alarm} = t \\ g^{+}[t] = 0 \\ g^{-}[t] = 0 \end{cases}$$

The algorithm takes three inputs: the first is the data x, the other two are *threshold* and *drift* values. There are many ways to implement the CUSUM algorithm. One of the ways is to find out positive ($g^{+}[t]$) and negative ($g^{-}[t]$) changes in the values of data (x) and then calculate its cumulative sum. The cumulative sum is then compared to the threshold value. If the cumulative sum at a given time t exceeds the threshold, it is made to start from zero, and an alarm(t) is generated to detect the change. It can be seen in the above Algorithm 1. The drift, which is another essential parameter, is used here to reduce the false positives. Therefore, the CUSUM algorithm mainly relies on tuning threshold and drift parameters.

According to Gustafsson (2000) [33], the following steps can be used for tuning purposes:

- Choose to begin with a high *threshold.*
- Pick a *drift* value that is half of the expected change or such that $g = 0$ more than half of the time.
- The next step would be to set a *threshold* to obtain detections.
- Decrease the *drift* to achieve faster detection
- Increase the *drift* to reduce false positives.
- The *drift* can also be increased in cases where the changes do not make sense.

The output of the algorithm gives an array of indexes where the changes are detected and their amplitude. This means it provides the location of the data frames in the form of an array to identify the attacks. The output also plots a graph of the detected changes and the data frames.

5 Empirical Evaluation

5.1 Performance Metrics

The performance evaluation of the proposed detection scheme is done by computing the detection rate (DR), the false positive rate (FPR) and the accuracy, which are classical metrics used for IDS evaluation. In addition to these two metrics, we are concerned also by the detection delay or mean-time-to-detect (MTTD), which is an important performance criterion when applying change-point detection.

Detection rate: The detection rate is defined as the number of attack samples detected by the model to the total number of attack samples in the data:

$$Detectionrate = \frac{Numberofattacksamplesdetected}{Totalnumberofattacksamples}\#$$

False-positive rate: The false-positive rate is defined as the ratio of the number of benign samples that have been reported as malicious to the total number of benign samples:

$$Falsepositiverate = \frac{Numberofbenignsamplesmislabelledasmalicious}{Totalnumberofbenignsamples}\#$$

Accuracy: computed as the ratio of correctly classified data points to total data points as follows:

$$Accuracy = \frac{True\ Positives + True\ Negatives}{True\ Positives + False\ Positives + False\ Negatives + True\ Negatives}$$

Mean absolute error rate: Different metrics are available to calculate the MTTD or time difference between the detection time for an attack and the actual time when the attack occurred. We use in this project the mean absolute error (MAE) metric to compute the MTTD:

$$MAE = \frac{\sum_{i=1}^{N} |Predicted(attack_i) - Actual(attack_i)|}{N}$$

where N is the total number attack instances.

The absolute value of the difference between the predicted and actual attack time is summed and normalized over each of the detected attack instances (or change points).

5.2 Evaluation Procedures

We run each of the algorithms on the datasets by varying the settings (i.e., set of algorithm parameter values). For each setting, we calculate the detection performance in terms of FPR, DR and MAE, and then we build a ROC curve plotting DR against FPR for different setting.

Since there is no training phase, the complete set of data are used for testing. For each setting, we run the algorithm separately on the benign and attack datasets.

In the proposed approach, once a change is detected, an estimate for its location is returned. If the observation at that location is malicious as per the original label, then we assume the change is caused by a malicious activity and the change is qualified as True Positive (TP). Otherwise, if the observation at the estimated location is normal, then the change is qualified as a False Positive (FP). If the detection algorithm fails to detect a malicious observation that might cause a change, we refer to the location of such an observation as a False Negative (FN). The remaining normal observations that do not cause any change are referred to as True Negatives (TN).

5.3 Evaluation Results

Figures 5 and 6 show the detection graph and ROC curve for the first Basic DoS attack. As seen in Table 4, the data throughput feature performs better than other features. The threshold and drift selected for attack are 0.82 and 0.1, respectively.

Figures 7 and 8 show the detection graph and ROC curve for Attack 2 (Broadcast DoS). Table 5 shows that the data throughput feature performs better than other features in this attack scenario. The threshold and drift selected for attack are 0.81 and 0.1, respectively.

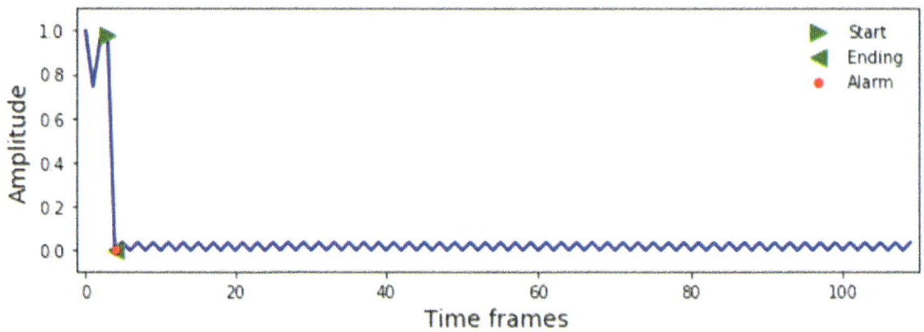

Fig. 5 CUSUM detection graph: Attack 1 (basic DoS) with data throughput feature

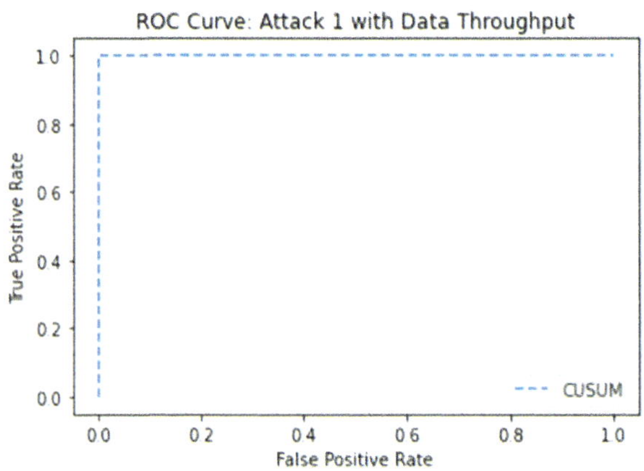

Fig. 6 ROC curve: Attack 1 with data throughput

Table 4 Performance evaluation: Attack 1

Attack 1: DoS attack (threshold: 0.82 and drift: 0.1)				
	Accuracy (%)	FPR (%)	TPR (%)	MAE (%)
Data throughput	100.0	0.0	100.0	0.0
Inter message gap	83.33	100.0	100.0	16.67
Bus utilization	85.71	0.0	50.0	14.29
Mode code	83.33	100.0	100.0	16.67

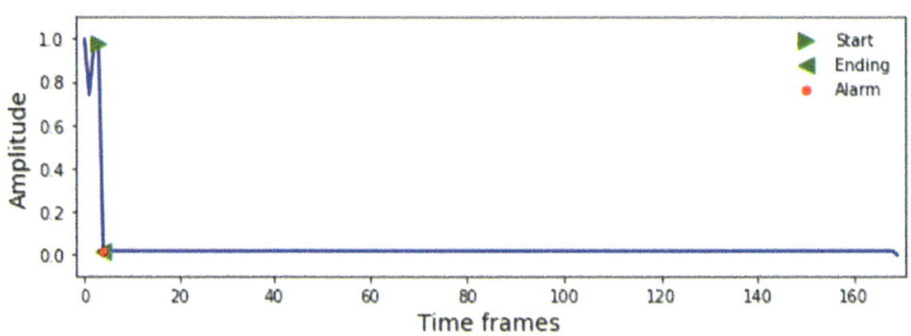

Fig. 7 CUSUM detection graph: Attack 2 (Broadcast DoS) with data throughput feature

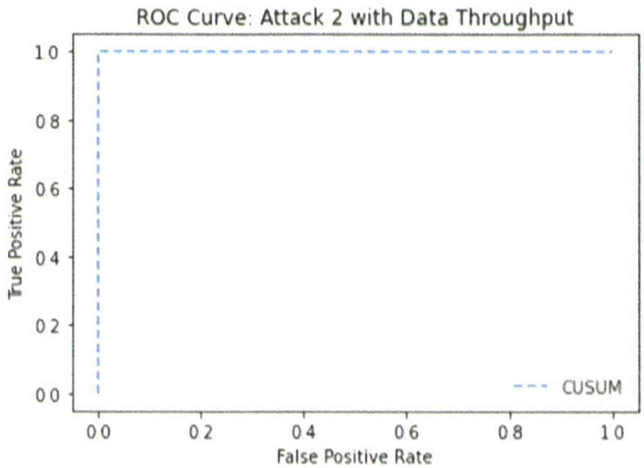

Fig. 8 ROC curve: Attack 2 with data throughput feature

Table 5 Performance evaluation: Attack 2

Attack 2: broadcast DoS attack (threshold: 0.81 and drift: 0.1)				
	Accuracy (%)	FPR (%)	TPR (%)	MAE (%)
Data throughput	100.0	0.0	100.0	0.0
Inter message gap	66.67	100.0	100.0	33.33
Bus utilization	80.0	10.0	100.0	20.0
Mode code	66.67	100.0	100.0	33.33

Inter message gap feature outperforms all other features in Attack 3 (Subtle injection), as seen in Table 6. The threshold and drift selected for attack are 0.9 and 0.1, respectively. Figures 9 and 10 show the detection graph and ROC curve, respectively.

Figures 11 and 12 show the detection graph and ROC curve for Attack 4 (Noisy injection). The threshold and drift selected for attack are 0.9 and 0.01, respectively. Table 7 shows that the inter message gap feature performs better than other features in this scenario.

Table 6 Performance evaluation: Attack 3

Attack 3: subtle injection attack (threshold: 0.9 and drift: 0.01)				
	Accuracy (%)	FPR (%)	TPR (%)	MAE (%)
Data throughput	59.29	39.639	0.0	40.71
Inter message gap	97.87	2.17	100.0	2.13
Bus utilization	45.39	55.07	66.67	54.61
Mode code	97.85	100.0	100.0	2.15

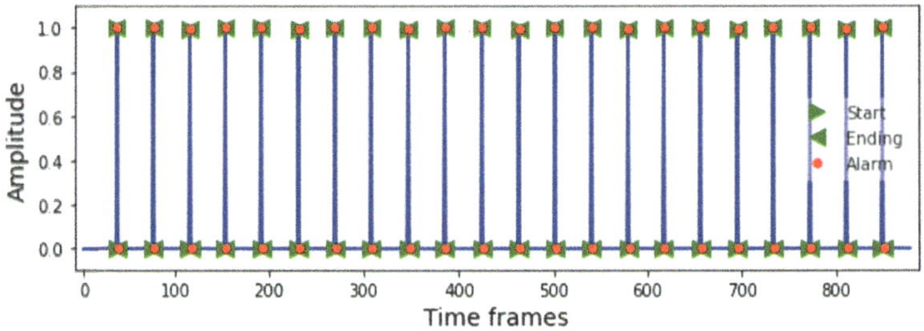

Fig. 9 CUSUM detection graph: Attack 3 (subtle injection) with inter message gap feature

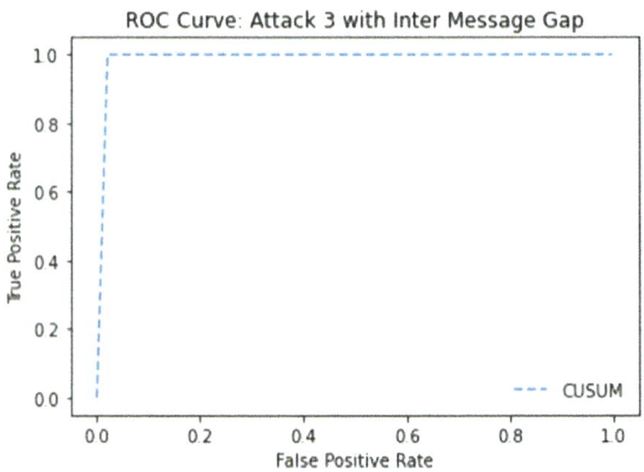

Fig. 10 ROC curve: Attack 3 with inter message gap feature

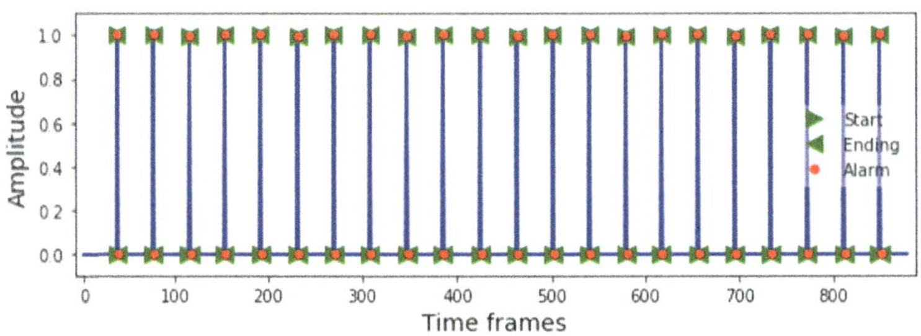

Fig. 11 CUSUM detection graph: Attack 4 (noisy injection) with Inter message gap feature

As seen in Table 8, the ModeCode feature outperforms all other features for Attack 5 (Logic attack). The threshold and drift selected for attack are 0.9 and 0.01, respectively. Figures 13 and 14 show the attack's detection graph and ROC curve.

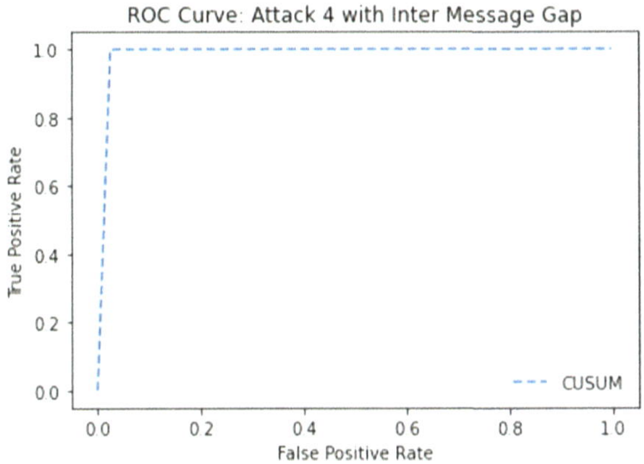

Fig. 12 ROC curve: Attack 4 with inter message gap feature

Table 7 Performance evaluation: Attack 4

Attack 4: noisy injection attack (threshold: 0.9 and drift: 0.01)				
	Accuracy (%)	FPR (%)	TPR (%)	MAE (%)
Data throughput	60.17	38.73	0	39.82
Inter message gap	97.50	2.56	100	2.50
Bus utilization	45.39	55.07	66.67	54.61
Mode code	97.87	100	100	2.13

Table 8 Performance evaluation: Attack 5

Attack 5: Logic attack (threshold: 0.9 and drift: 0.01)				
	Accuracy (%)	FPR (%)	TPR (%)	MAE (%)
Data throughput	53.91	41.23	27.78	46.09
INTER MESSAGE GAP	92.35	5.20	0.0	7.65
Bus utilization	52.45	51.67	73.91	47.55
Mode code	84.62	18.33	100.0	15.38

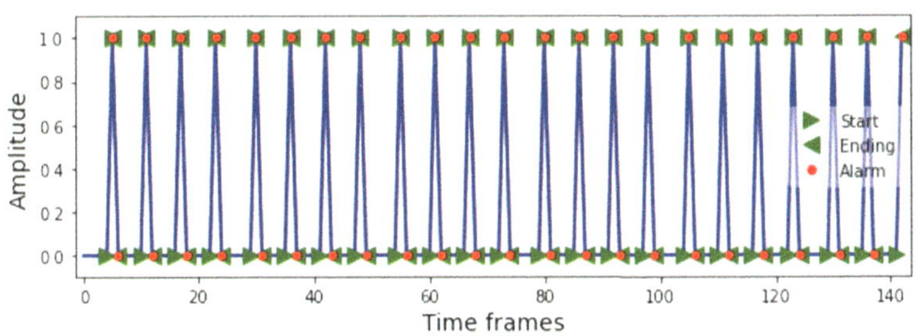

Fig. 13 CUSUM detection graph: Attack 5(Logic) with ModeCode feature

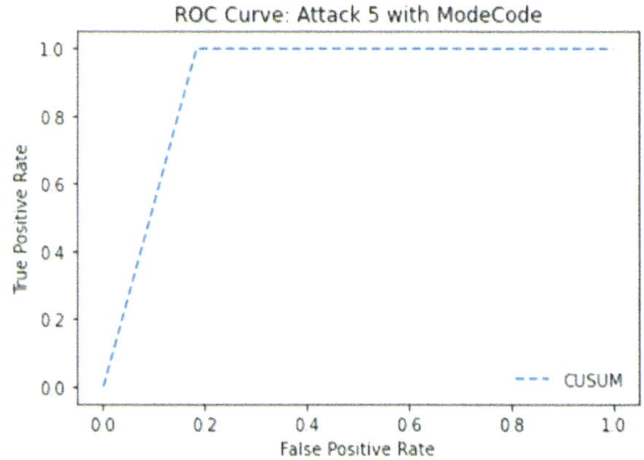

Fig. 14 ROC Curve: Attack 5 with ModeCode feature

Figures 15 and 16 show the detection graph and ROC curve for Attack 6 (Combination attack). Table 9 shows that the inter message gap feature performs better than other features in this attack scenario. The threshold and drift selected for attack are 0.9 and 0.01, respectively.

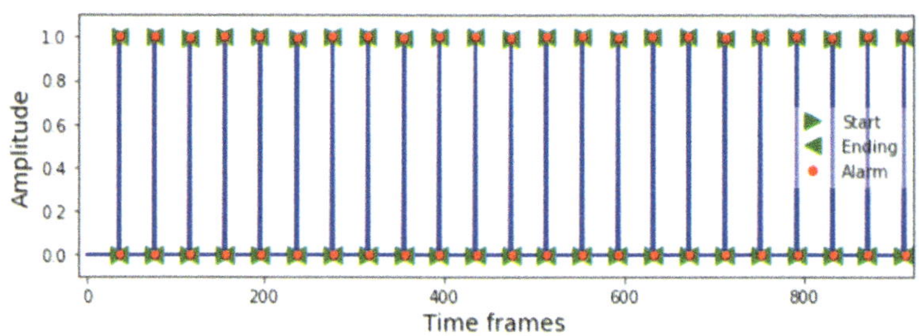

Fig. 15 CUSUM detection graph: Attack 6 (combination) with Inter message gap feature

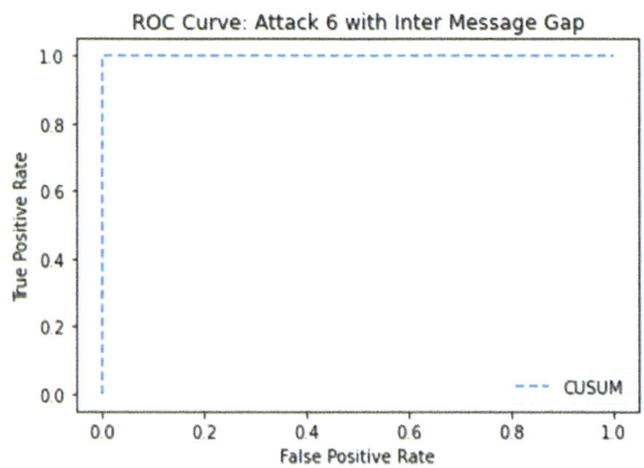

Fig. 16 ROC curve: Attack 6 with Inter message gap feature

Table 9 Performance evaluation: Attack 6

Attack 6: combination attack (threshold: 0.9 and drift: 0.01)				
	Accuracy (%)	FPR (%)	TPR (%)	MAE (%)
Data throughput	93.04	3.6	0.0	6.96
Inter message gap	100.0	0.0	100.0	0.0
Bus utilization	56.64	40.87	0.0	43.36
Mode code	68.53	30.65	50.0	31.47

5.4 Results Discussion

The above results show that the data throughput will help detect attacks like Denial-of-Service (DoS) attacks. The basic DoS is carried by a rogue terminal sending random words to a specific RT. While, in the broadcast DoS attack, the rogue terminal sends broadcast messages to all other remote terminals. The inter message gap feature is better suited in attacks where fake data injection occurs. The rogue terminal manipulates either a property value or the entire message while sending the data. It is evident from the above findings that the subtle injection attack where rogue terminal changes one of the message properties and the noisy injection attack where the rogue terminal changes entire data are detected with high accuracy and low FPR. Logic attacks are detected efficiently by the mode code feature. These attacks are carried out by sending random commands like shutdown in the middle of regular communication. The last attack, a combination of logic and fake data injection, was detected accurately with the Inter message gap property.

These findings show that the CUSUM algorithm's primary objective is to detect the attacks as efficiently as possible. However, to build a complete IDS, the above results should use a certain combination of features to detect attacks. This can be done by combining more than one feature to detect the anomaly using logical operators. The performance properties like accuracy, FPR, TPR from the output from the algorithm can be used to implement a logic further and make the technique more robust. So, this work brings us a step closer to building a complete IDS for the MIL-STD-1553 standard.

6 Conclusion

MIL-STD-1553 is a widely accepted military standard across various platforms worldwide. However, it has many attack vectors associated with it and could have devastating results if exploited. Our work aimed to develop an intrusion detection technique for the MIL-STD-1553 standard using the CUSUM algorithm for change detection.

This aim was achieved by using the timing properties of the protocol and extracting four features based on it. Then, a CUSUM algorithm was implemented to use the extracted features and detect six attacks on the MIL-STD-1553 communication bus. Finally, performances and tests were carried out to validate the algorithm, and the results show that individual features gave very positive results to detect specific attacks. This proves that the standard's time-based properties can effectively detect such intrusions.

Future work on this could include using additional properties like periodicity and response time, which have proven promising in the past [2]. In addition, future work would include implementing some logic based on the output of the features to identify attacks and make the technique more robust. This would include working on the performance parameters, applying a combination logic using the logic operators (e.g., AND, OR etc.), and using the additional properties discussed earlier. Though the technique presented in this project successfully detects six attacks, it does not cover all attacks, and more work can be done to include newer attack vectors for the bus. Also, the CUSUM technique used in this project brings us a step closer to building a complete IDS for the MIL-STD-1553 standard.

References

1. Edition S Mil-std-1553 designer's guide. Available online: http://www.ddc-web.com
2. Losier B (2019) Design of a time-based intrusion detection algorithm for the MIL-STD-1553. Master Thesis, Royal Military College of Canada, Kingston, Ontario, Canada
3. McGraw RM, Fowler MJ, Umphress D, MacDonald RA (2014) Cyber threat impact assessment and analysis for space vehicle architectures. In: SPIE Defense+ Security (2014), International Society for Optics and Photonics, pp. 90850K–90850K
4. Huitsing P, Chandia R, Papa M, Shenoi S (2008) Attack taxonomies for the Modbus protocols. Int J Crit Infrastruct Prot 1:37–44. https://doi.org/10.1016/j.ijcip.2008.08.003
5. Hayden PM, Woolrich DK, Sobolewski KD (2017) Providing cyber situational awareness on defense platform networks. Cyber Defense Rev 2(2):125–140.
6. Anderson JP (1980) Computer security threat monitoring and surveillance. Technical report, James P Anderson Co., Fort Washington, Pennsylvania
7. Denning D (1987) An intrusion-detection model. IEEE Trans Softw Eng SE-13(2):222–232
8. Aldribi A, Traore I, Moa B, Nwamuo O (2020) Hypervisor-based cloud intrusion detection through online multivariate statistical change tracking. Comput Secur 88
9. Stan O, Elovici Y, Shabtai A, Shugol G, Tikochinski R, Kur S (2017, July 17) Protecting military avionics platforms from attacks on MIL-STD-1553 communication bus. arXiv:1707.05032v1 [cs.CR]
10. Yahalom R, Barishev D, Steren A, Nameri Y, Roytman M, Porgador A, Elovici Y (2019) RT spoofing attacks on MIL-STD-1553 communication traffic. Mendeley Data v3, Elsevier. https://doi.org/10.17632/jvgdrmjvs3.3
11. Nguyen TD (2015) Towards mil-std-1553b covert channel analysis. Tech. rep., Monterey, California. Naval Postgraduate School
12. Lu W, Traore I (2005) A new unsupervised anomaly detection framework for detecting network attacks in real-time. In: 4th international conference on cryptology and network security (CANS), Xiamen, Fujian Province, China, 14–16 December, 2005, Lecture notes in computer science, vol 3810, pp 96–109. Springer, Berlin, ISBN 3-540-30849-0 [Y.G Desmeddt et al. (eds)]
13. Ferragut E, Laska J, Bridges R (2012) A new, principled approach to anomaly detection. In Proceedings—2012 11th international conference on machine learning and applications, ICMLA 2012, vol 2, pp 210–215. https://doi.org/10.1109/ICMLA.2012.151

14. Yousef W, Traore I, Briguglio W (2021) UN-AVOIDS: unsupervised and nonparametric approach for visualizing outliers and invariant detection scoring. Trans Inf Forensics Secur 16:5195–5210
15. Page ES (1954) Continuous inspection scheme. Biometrika 41(1/2):100–115
16. Grigg OA, Farewell VT; Spiegelhalter DJ et al (2003) The use of risk-adjusted CUSUM and RSPRT charts for monitoring in medical contexts. Stat Methods Med Res 12(2):147–170
17. James NA, Matteson DS (2013) ecp: an r package for nonparametric multiple change point analysis of multivariate data. arXiv preprint arXiv:1309.3295
18. Matteson DS, James NA (2014) A nonparametric approach for multiple change point analysis of multivariate data. J Am Stat Assoc 109:334–345
19. Tartakovsky AG, Rozovskii BL, Blazek RB, Kim H (2006) A novel approach to detection of intrusions in computer networks via adaptive sequential and batch-sequential change-point detection methods. IEEE Trans Signal Process 54(9):3372–3382
20. Akoglu L, Faloutsos C (2010) Event detection in time series of mobile communication graphs. In: Army science conference, pp 77–79
21. Sequeira K, Zaki M (2002) Admit: anomaly-based data mining for intrusions. In: Proceedings of the eighth ACM SIGKDD international conference on Knowledge discovery and data mining. ACM, pp 386–395
22. Wang H, Zhang D, Shin KG (2004) Change-point monitoring for the detection of dos attacks. IEEE Trans Dependable Secure Comput 1(4):193–208
23. Hawkins DM, Qiu P, Kang CW (2003) The changepoint model for statistical process control. J Qual Technol 35(4):355
24. Hawkins DM, Zamba K (2005) A change-point model for a shift in variance. J Qual Technol 37(1):21
25. Zamba K, Hawkins DM (2006) A multivariate change-point model for statistical process control. Technometrics 48(4):539–549
26. Ross GJ, Adams NM (2012) Two nonparametric control charts for detecting arbitrary distribution changes. J Qual Technol 44(2):102
27. Ross GJ, Tasoulis DK, Adams NM (2011) Nonparametric monitoring of data streams for changes in location and scale. Technometrics 53(4):379–389
28. Hinkley DV, Hinkley EA (1970) Inference about the change-point in a sequence of binomial variables. Biometrika 57(3):477–488
29. Stephens D (1994) Bayesian retrospective multiple-changepoint identification. Appl Stat 159–178
30. Page ES (1954) Continuous inspection schemes. Biometrika 41:100
31. Kawahara Y, Sugiyama M (2012) Sequential change-point detection based on direct density-ratio estimation. Stat Anal Data Min 5:114–127
32. Duarte M (2021, April 3) CUSUM detection. [Online]. Available https://nbviewer.org/github/demotu/detecta/blob/master/docs/detect_cusum.ipynb. Accessed 13 Feb 2022
33. Gustafsson F (2000) Adaptive filtering and change detection. Wiley, Chichester, England, New York

Secure Design of Cyber-Physical Systems at the Radio Frequency Level: Machine and Deep Learning-Driven Approaches, Challenges and Opportunities

Ceren Comert, Omer Melih Gul, Michel Kulhandjian, Azzedine Touazi, Cliff Ellement, Burak Kantarci, and Claude D'Amours

Abstract With the deployment of new 5G services, many of the critical infrastructures such as connected vehicles, remote healthcare and smart infrastructures will be deployed on radio frequency (RF)- based networks. As such, society will be heavily dependent on the ability to protect these new wireless networks as well as the radio spectrum. Solutions such as artificial intelligence (AI)-based transmitter fingerprinting to identify and track unintended interference sources or malicious actors will be one of the several key technologies required to meet the needs of the next generation wireless networks as this technology is deployed as part of a critical infrastructure (CI). As an example, connected and autonomous vehicles (CAVs) can be considered under these cyber-physical systems and critical infrastructures. As 95 percent of new automobiles are expected to be equipped with vehicle to infrastructure (V2I), vehicle to vehicle (V2V), and other telecommunications capabilities by 2022. To ensure the safety of the public, new and automated techniques are needed to protect CAVs on the road from unintentional or malicious interference. Against these requirements,

C. Comert (✉) · O. M. Gul · B. Kantarci
University of Ottawa-Kanata North Campus. 200, 535 Legget Drive, Ottawa, ON K2K 3B8, Canada
e-mail: ccomert@uottawa.ca

O. M. Gul
e-mail: ogul@uottawa.ca

B. Kantarci
e-mail: burak.kantarci@uottawa.ca

M. Kulhandjian · C. D'Amours
University of Ottawa, 800 King Edward Avenue, Ottawa, ON, Canada
e-mail: mkulhand@uottawa.ca

C. D'Amours
e-mail: cdamours@uottawa.ca

A. Touazi · C. Ellement
ThinkRF, 390 March Rd, Kanata, ON K2K 0G7, Canada
e-mail: azzedine.touazi@thinkrf.com

C. Ellement
e-mail: cliff.ellement@thinkrf.com

I. Traore et al. (eds.), *Artificial Intelligence for Cyber-Physical Systems Hardening*, Engineering Cyber-Physical Systems and Critical Infrastructures 2,
https://doi.org/10.1007/978-3-031-16237-4_6

this chapter presents the state of the art in real time decision support systems for the cyber-physical systems that build on critical infrastructures such as CAVs, through radio fingerprinting solutions. In this chapter, we first present the legacy approaches used to detect, classify and identify a transmitter, and then we move towards the machine and deep learning-based (ML/DL) approaches for transmitter identification using RF fingerprinting techniques. Following upon a comparative study on the open issues, challenges, and opportunities towards ML/DL-driven security of the critical cyber-physical systems through RF fingerprinting.

Keywords 5G · Cyber-physical system · Radio frequency fingerprinting · Machine learning · Deep learning · Artificial intelligence · Transmitter fingerprinting · Connected and autonomous vehicle · Critical infrastructure · Interference detection · Classification · Radio frequency security · Authentication · Intrusion detection · Anti-spoofing · Dataset

1 Introduction to Cyber-Physical Systems

Cyber-physical systems (CPS) are the systems that consist of cyber and physical components. Having an embedded network, cyber-physical systems have the ability to interact with the physical environment. They also have the capacity to adapt to the changing environment and to control it if required [9].

A CPS has a control unit, sensors and actuators. These are needed to interact with the surrounding environment. A communications interface is also needed to communicate with other systems. The cyber part of the CPS is responsible for calculating and sending inputs to the actuators. Data for the cyber part is obtained by the sensors. Then, actuators can take necessary action in the physical environment [72].

When linked to the Internet with a central (cloud) processing node and edge nodes, a CPS can be called as the Internet-of-Things (IoT) [44]. Another concept is the web of things (WoT) which builds on top of IoT. This is achieved by merging networked things into the Web. Consequently, WoT can use web-based protocols like uniform resource identifier (URI), hypertext transfer protocol (HTTP) etc. As IoT connects things to the Internet, WoT connects the physical devices to the Web. The WoT framework as explained in [20] has layers such as WoT device, kernel, overlay, content and application programming interface (API). Moreover a CPS interface connects WoT to the physical world.

Depending on their implementation complexity, CPS can be low-end or high-end. Low-end implementations are linearly complex, distributed and networked. They usually have feedback controller systems. High-end ones, on the other hand, are non-linearly complex, and decentralized. They are commonly intelligent systems with learning capability. If used in critical infrastructures such as nuclear or power plants, autopilot systems and autonomous vehicles, a malfunctioning CPS can pose a risk to property as well public safety. Moreover, such operations generally have uncertainty, which creates challenges for dependability and maintenance [39].

Additionally, the physical environment is not completely predictable and observable. CPSs also need to function in uncontrolled environments. Therefore, the design of CPSs must be robust [55]. The CPS Framework includes aspects and facets which should be considered when designing such systems. Aspects can be referred to as main categories of concern. Facets are perspectives of bounding responsibilities in the engineering process [34]. Facets can be listed as follows:

- Conceptualization: The activities that output a model of a CPS. It is used for defining what should be included.
- Realization: The term used to describe how the CPS should function. Trade-offs and designs can be given as example.
- Assurance: A desired level of confidence and methods for reaching this level of confidence are defined.

Aspects of CPS can be summarized as follows:

- Functional Aspects: represents functionality and actuators including sensing of physical environment
- Business Aspects: is mainly about enterprise, price and marketing.
- Human Aspects: represents human interactions with systems.
- Trustworthiness: is concerned with, among other things, security, privacy and safety of the CPS.
- Timing: represents the concerns associated with time and frequency within the CPS. Latency and timestamp are main examples.
- Boundaries: stand for topological or functional limitations.
- Composition Aspects: are relevant to the component assembly.
- Life cycle: represents concerns on lifetime of CPS.

When aspects and facets are taken into consideration, CPSs should also have requirements [42] that can be listed as follows.

- Architecture
- Middle-ware design
- Quality of service
- Control
- Real time management
- Systems security

This book chapter is structured as follows. The rest of this section provides information about cyber-physical systems along with their application areas and security methods. Section 2 focuses on critical infrastructures. Section 3 continues with the fundamentals of RF fingerprinting method which is a promising solution for enhancing security. Section 4 contains information about security precautions including RF fingerprinting methods. The chapter continues with deep and machine learning methods for RF fingerprinting in Sect. 5. Related literature is included in this section to provide more knowledge. Section 6 provides information about open issues and challenges. The chapter is concluded with Sect. 7.

1.1 Security and Application Areas of CPS

The applications of CPS are vast. Some examples are concerned with, among other things, security, privacy and safety of the CPS: power grids, smart home applications, health care systems, assisted living, medical monitoring, wearable devices, agricultural systems, military systems, meteorology, traffic control, energy conservation, industrial control systems and autonomous systems (such as drones, autonomous or self driving vehicles etc.), mobile and communications systems [3, 45, 52, 85].

Figure 1 illustrates a few examples of the application areas of CPSs.

The application areas listed above are broad and are made up of sub-areas. For example, health care systems include, among other things, electronic patient records, home care, and assisted living. However, medical CPS are often limited to health requirements. To illustrate, consider the X-Ray example of [74] where the CPS is used to disable the patient's respirator before the X-Ray is taken to ensure the safety of the patient.

One of the most commonly studied and important application areas of CPS is electrical power grids. One drawback in this area is balancing the supply with respect

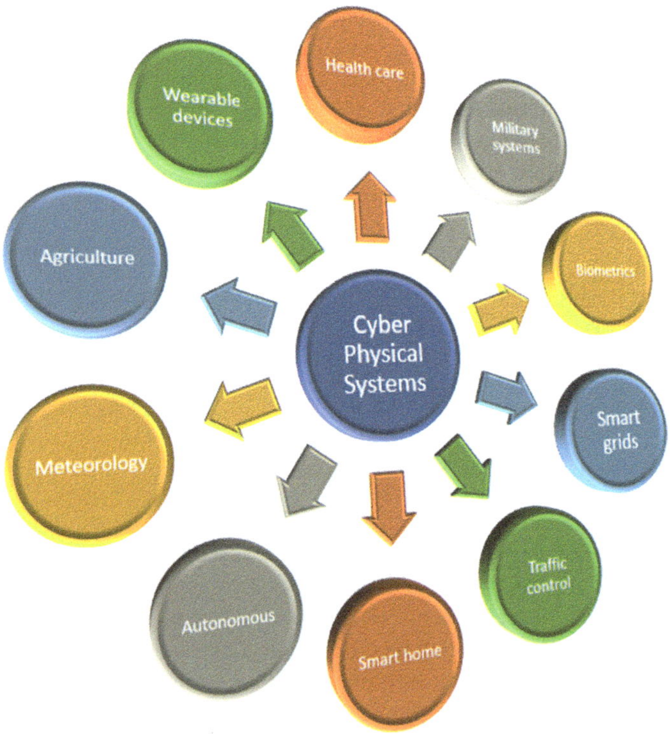

Fig. 1 Application areas of cyber-physical systems

to the individual users' consumption demand. Likewise, outputs of renewable energy sources can fluctuate significantly and should be predicted correctly. Therefore, the balance between supply and the demand is a challenging problem.

Another application area of CPS is wireless sensor networks (WSN). WSNs have applications on CPSs such as smart monitoring, cyberspace and border security tracking, more importantly for gathering information about environments. With the advances in WSNs, CPS might be applied as a solution for connected and autonomous vehicles (CAVs). Intelligent traffic control can be possible. However, WSNs have several challenges in terms of energy, space and time.

While traditional sensor networks are usually designed for some specific tasks and have a few nodes connected to a central processor. Differently, WSNs have many nodes which are capable of sensing and measuring the surrounding environment. With a use of many nodes, line-of-sight (LOS) requirements can be achieved. It should be noted that WSN applications on CPS for communications don't have existing predefined infrastructures [67].

Varying sensor types can be deployed according to the purpose of use, such as electric, light, acoustic, pressure and humidity etc. Smart sensors use two way communication and logic circuits to calculate and process information. Smart sensing networks are used in applications like smart home and smart cities. Distributed WSNs and mobile ad-hoc networks (MANETs) are also commonly used in CPS applications, especially smart health care, rescue and transportation systems.

CPS are also used together with the term of *Industry 4.0*. As described in [19, 44] the first revolution was *mechanization* followed by the *mass production*. Third one is the *digitization* with the electronics and Internet. CPS is counted as the fourth revolution which names the term Industry 4.0. Some technologies for Industry 4.0 can be listed as additive and smart manufacturing, augmented reality, autonomy, big data and cloud applications and so on [86].

Becoming a part of our daily lives, CPS provides us many advantages. However, rising usage and dependency of CPS can lead to some security threats that should be considered [9].

An attacker can breach *confidentiality* by eavesdropping on communications between sensors, controllers, and actuators. *Integrity* is affected if these communications can be altered by the attacker. Cyber as well as wireless attacks can prevent CPS availability, by causing failures in the computer technology or by jamming communications, etc.

Two other important aspects when faced with the possibility of attacks are *reliability and authentication*. Authentication can be defined as the need for confirming that involved users are legitimate. While designing a CPS network, *nonrepudiation* should be provided by proving the actions in order to take precautions.

Some common security threats within the Wireless domain include spoofing, denial of service (DoS), data tampering and corruption, and injection of false information. Attacks on computer-based systems are also possible with trojans, viruses and worms. Since cyber-physical systems interact with physical world, actuators can also be prone to attacks such as injection of false radar signals, global positioning system (GPS) spoofing, and interference with cameras. Communication systems of cyber-physical systems may be prone to attacks such as sybil attacks, selective forwarding, packet spoofing and replaying etc.

Security problems in cyber-physical systems have become a world-wide concern, thus, the design of secure systems is an important and contemporary area of research. As security attacks continue, the demand for new measures to protect CPS will persist. These measures can be defined as cyber-defense process that includes a number of fundamental steps, namely prevention, detection, reaction, and forensics [83].

1. Prevention: This stage can be explained as monitoring the systems and detecting its vulnerabilities. An example of prevention can be given as access control.

Access control systems (ACSs) are used for managing privileges to guarantee secure access. Firewalls (FWs) are one of the important concepts of prevention. They can analyze packet content, detect some malicious content, deny unauthorized connections etc. The CPS uses internal firewall system that prevents attackers from connecting to server outside of the network [62].

Intrusion prevention systems (IPSs) method is highly important. To monitor traffic flows, risk assessment techniques can be used to scan vulnerabilities and threats, and enumerate them [83].

2. Detection: This stage includes detecting and reporting the events. For this stage, intrusion detection systems (IDS) are used most commonly. IDS can be divided into two groups based on their detection method.

Anomaly based IDS attempts to identify deviations from normal system activity traces and traffic. Signature based IDS performs the detection by matching some rules to uncover known attack patterns. Additionally, firewalls are also considered as IDSs, too.

3. Reaction: After detection stage, this stage takes place. Reaction stage includes responding and blocking ongoing attacks and, providing effective precautions. At this stage, the system is returned into its normal state.

4. Forensics: stage manages investigating events after the attacks are addressed. Thus, parties can understand mistakes and avoid them in the future.

2 Critical Infrastructures

Since CPS is used in critical infrastructure, an additional section is included about CIs. The term 'critical infrastructure' can be explained as critical services or industries that play crucial role for a particular business or for the public. A CI is called critical if its damage or failure of operation causes a vital impact on health, economics and safety. A loss of CIs can have serious effect on society, and it might also cascade to other CIs.

CI includes water supply facilities (including distribution, transport, wastewater, storage, treatment), gas production (including transport, distribution), oil products (including transport, distribution, production), telecommunication, electricity generation (including distribution, transmission), which are required for society and the economy to function properly [42].

An important concept is the notion of inter-dependency, which can be explained as the relation between infrastructures in which each of them is influenced by the other. An example is the inter-dependency between electric power grids and gas systems outlined in [80]. Any failure in the electrical system might cascade to the natural gas system.

CIs can be considered as interdependent with other infrastructures if the following conditions hold [61].

- Input: The CI needs one or more inputs from another CI.
- Shared: Some components of the CI are shared with other CIs. For example electric power and network [26].
- Exclusive-or: One CI or another CI can be in use for the service. Simply, both CIs cannot be in use at the same time. As an example, limiting a usage of power generator for CI.
- Co-location: At least two CIs' components are co-located in the same region. The location of the CI faces the risk of being damaged if a disaster occurs in the area.

Methods for studying inter-dependencies of critical infrastructures can be categorized into three groups [92]. These are conceptual, model and empirical based approaches. Simulation and modeling approaches study infrastructure models to analyze disturbances and their effects on such systems. Whereas empirical approaches analyze data and statistics obtained from the actual events. Though they can be more effective than simulation approaches due to being based on actual events, empirical methods cannot obtain information about the events that haven't occurred. Conceptual inter-dependencies studied in [80]:

- Physical inter-dependency: Which can be explained as coupling between the systems inputs and outputs. Some other infrastructure might need to use a commodity created by another
- Cyber inter-dependency: Concept that is used to explain the dependency on the information infrastructure
- Geographical inter-dependency: Some events like geographical hazards can create damage or disturbances to CIs
- Logical inter-dependency: This term is often used for infrastructures having reciprocal effects without any above-mentioned inter-dependencies

Figure 2 demonstrates an example of inter-dependencies of critical infrastructures.

Critical infrastructures are also complex systems. Complex systems have a huge number of dimensions, non-linear attributes, and powerful interactions. Their components are heterogeneous. Heterogeneity means the differences in the elements and connections. In power grids, powerful heterogeneity occurs since facilities are connected by centralized systems to consumer branches. Heterogeneity creates dynamic and structural complexity. Most well known example for the complex system that can be given is the Internet. Needless to say, the Internet is broadly used by many people and companies [108].

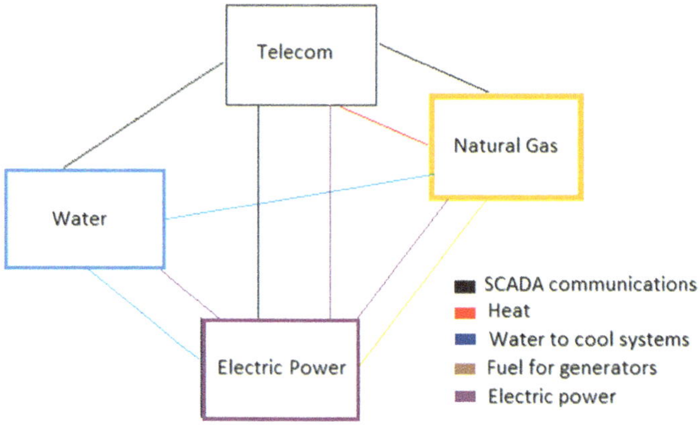

Fig. 2 Interdependencies of critical infrastructures

Critical infrastructures can be monitored and secured such as IDS, anti-spoofing techniques etc. RF fingerprinting can be used to secure CPS by detecting devices unique fingerprints. The chapter continues by explaining some fundamentals about RF fingerprinting.

3 Fundamentals of Radio Frequency Fingerprinting

RF fingerprinting is a process of detecting unique characteristic of transmitters [98].

RF fingerprinting can be used for detecting the unique radio transmitters at the physical layer level. RF fingerprinting shows the distinguishing characteristics of the transmitters' components, and their properties [18]. RF fingerprinting is a very promising method for securing cyber-physical systems.

Because of some hardware imperfections in manufacturing process, specific characteristics of the transmitter are presented in the transmitted signal, such as imbalances across a number of signal parameters. As these imperfections are unique, they can be used for detecting specific transmitters to secure communication network.

RF fingerprint feature extraction can rely on different features, for example, in phase and quadrature imbalance (I and Q), DC offset, power amplifier non-linearity, differential non-linearity, carrier frequency offset, clock offset, phase shift difference, etc. [13, 107].

The features that can be obtained from the received RF waveform can be classified as location dependent and location independent (radiometric) features [100].

Most commonly known location dependent features are received signal strength (RSS) and channel state information (CSI).

RSS technique depends on the average signal power at the receiver side, attenuation and power of transmission. Even the same transmitters which are at different

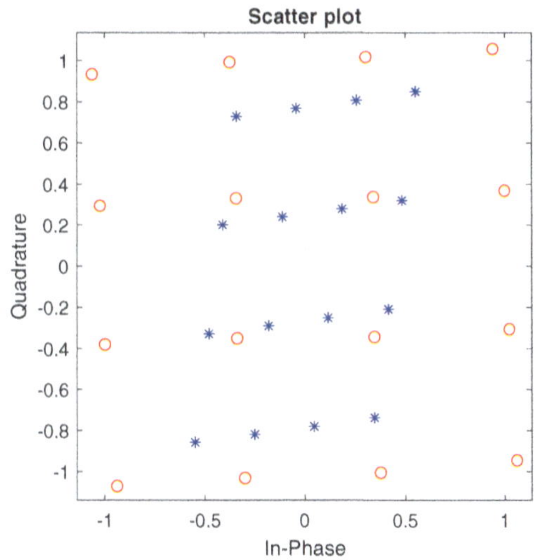

Fig. 3 Power amplifier non-linearity

locations may have different RSS. That is why, this method is classified as location dependent. However, transmitters at close locations may have close RSS values. This method might not be effective if a transmitter is aimed for identification at different locations.

CSI indicates known information about channel assets, and it contains merged effects of fading, scaling and so on. CSI at a receiver side may change significantly when the receiver is moved. This change often occurs because of small scale fading. Location independent features refer to imperfections of individual devices. Imperfections caused by the manufacturing process that are insignificant and do not impact communication specifications, are typically large enough to be used in detecting specific transmitters. Developed by Brick et al., passive radiometric device identification (PARADIS) is designed for using magnitude and phase errors as features [100].

Location independent features can be categorized as waveform and modulation based. Waveform techniques focus on time and frequency representations, while modulation techniques focus on I/Q symbols. In the following, more detailed knowledge are provided about the above mentioned features.

PA Non-linearity: Even though PAs are commonly linear, they saturate in high voltages. PA non-linearity has effects on communication. This non-linearity can be extracted and used as a feature [13]. Figure 3 shows PA distortions in constellation figures.

Phase Shift Difference: Crystal oscillators creates sine waves for up-link conversation [105]. Both the carrier frequency offset and phase noise introduce a phase shift to the constellations.

Carrier Frequency Offset: Carrier frequency offset occurs because of frequency mismatch in crystal oscillators. It can be used as a feature when identifying specific transmitters.

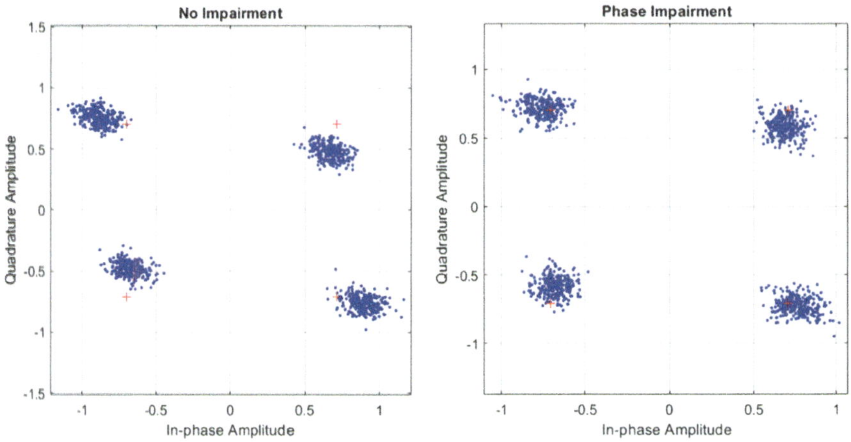

No Impairment Phase Impairment

Fig. 4 IQ Imbalance

Differential Non-linearity: Differential non-linearity is caused by the discrepancy between ideal values and obtained analog values of digital codes [13].

I/Q Imbalance: Complex down converter multiplies the signal by a complex wave. In order to perform such complex conversation, both sine and cosine waves are needed. By using both, the receiver is divided into in phase (I) and quadrature (Q) branches.

For communication, these branches need to have the same amplitude with ninety degrees of phase difference. The mismatch between I and Q branches causes IQ imbalance [57].

IQ imbalance can be observed more clearly in the frequency domain. Hardware imperfections and tolerances of the components (like imperfect splitting ratio of couplers, and polarization splitters) can cause amplitude and phase imbalance [66]. Figure 4 demonstrates IQ imbalance in constellation figures.

4 Application Areas of RF Security

Most common applications areas of RF security include authentication, geo-location and tracking, intrusion detection, RF interference detection and anti-spoofing. These application areas of RF security are presented in more detail in the rest of this section.

4.1 Authentication

The state of the art offers various methods to authenticate vehicles. In [94], the proposed model stores sensitive information temporarily and deletes them regularly. By this way, this sensitive information is being disposed before being obtained by adversaries. The authors mention that commonly used conditional authentication model, conditional privacy-preserving authentication (CPPA) relies on a tamper proof device (TPD). However, TPDs are not feasible due to practical security requirements and are sensitive to vehicle shocks which may cause information loss as a result of attacks. Therefore, a more realistic and practical model is needed.

In this framework the key used for authentication is generated by the vehicle itself and stored in the TPD so as to prevent impersonation attacks. The framework also uses authentication via trusted authority. The design in [94] enables authentication without real time interactions. Usage of pseudo-identities is another concept to keep identities safe. A trusted authority (TA) can verify identities. Additionally, since only a TA can verify identities, unwanted tracking of vehicles is also prevented. This model is designed against replay, impersonation and modification attacks.

Another authentication framework named signature and prediction TESLA for vehicular ad-hoc network (VANETs) is proposed in [58]. This framework uses a method for authentication that merges elliptic curve digital signature and TESLA algorithms. The TESLA model uses symmetric cryptography for authentication. It is not strictly synchronized, but requires time upper bound. Recommended by the dedicated short range communications standard such as IEEE801.11p, vehicles broadcast a safety message also known as beacon containing important information such as location and route. An attacker can perform DoS attack easily by sending fake information.

The TESLA model provides an alternative to signatures by using symmetric cryptography. It is resilient against DoS attacks in V2V communications. Nevertheless, a drawback of TESLA is that the receiver has to buffer packets during one disclosure delay before it can authenticate them. This situation makes TESLA impractical when instant identification is needed. As a result a more practical scheme was designed by merging the TESLA model with Merkle hash tree (MHT). The steps taken for this model include key generation, beacon prediction, building MHT and broadcasting signatures.

RF fingerprinting is based on detecting and understanding the unique RF characteristics of a transmitter created by its hardware imperfections. When RF fingerprinting is used to verify transmitter authenticity, RF characteristics such as I/Q imbalance can be used to identity spoofing and thus making it challenging for attackers. Therefore, RF fingerprinting can also be used to authenticate legitimate devices. Additionally, authentication can be more robust since it is based on radio frequency-level signatures [43]. As an example, Alladi et al. designed an authentication scheme based on physical unclonable functions (PUFs) [4]. The research in [10] introduces device authentication codes that integrate RF fingerprinting into IoT device authentication.

4.2 Geo-location and Tracking

GPS was the first reliable system for localization. Even though GPS can be successful in some applications, it may not be effective indoors and in urban areas. This and GPS-denied areas revealed a need for new geo-location techniques to be developed. IEEE 802.11 standard was introduced as a cellular localization technique, and RF fingerprinting can be used for outdoor or indoor localization [30]. Moreover, confirmed emitters can be tracked with RF fingerprinting. Channel impulse response (CIR)-based fingerprinting and frequency channel function (FCF)-based techniques can be used for localization.

In parametric geo-location techniques, the location system collects parametric information (as explained below) about a transmitting device to estimate its location [65].

- Times of arrival (TOA) method can estimate position when three reference points are available.
- Time differences of arrival (TDOA) method can estimate position when three reference points are available. This method uses time differences of intersections of the hyperbole.
- Angles of arrival (AOA) method can estimate position when two reference points to the mobile device are available [30].

The geo-location technique is formed from the idea that the signals at different locations will have different characteristics from each other [41]. To illustrate, mobile device's location can be found with this technique.

4.3 Intrusion Detection

Deep and machine learning methods can be used in intrusion detection. As an example, in [15], a machine learning based IDS for Cloud environments is introduced. To develop a network IDS based on neural network a hybrid model is used. Genetic algorithms are merged with *parallel processing* and optimization methodologies such as *fitness values hashing*. In [56], signature based, battery based and agent based methods are compared. Additionally, pattern classification methods are investigated. Some works for mobile systems are surveyed; as an example, in [11], behavior based IDS for mobile systems is developed.

RF fingerprinting can be a strong tool to detect intrusions. It can also identify imitation attacks. Attackers can perform replay attacks by mimicking legitimate users. Similar to the traditional fingerprints methods, RF fingerprinting can pave the way for the security systems to identify attacks [43].

For example, in order to detect medium access control (MAC) spoofing attacks an anomaly based IDS is introduced in [37] by leveraging RF fingerprinting with T-square statistical method.

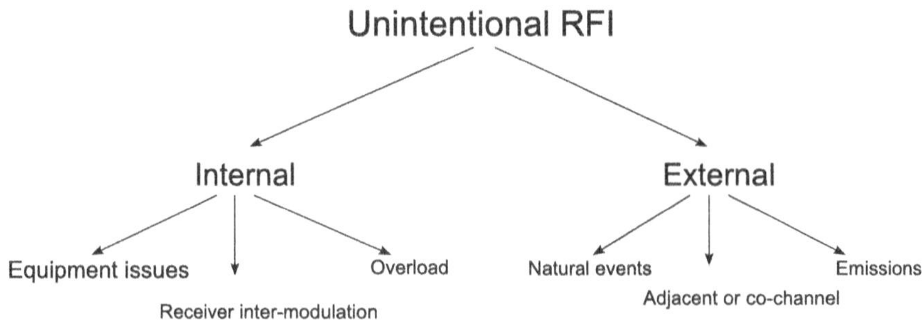

Fig. 5 Unintentional RFI

4.4 Interference Detection and Traditional Approaches

Radio frequency interference (RFI) that affects the system could be intentional or unintentional. Intentional RFI includes illegal sources such as jamming attacks. As jamming can be used for malicious intents, it can also be used for Wi-Fi, GPS jamming for workplaces or military purposes.

Unintentional interference is caused by outdated, degraded devices. Chargers, relays and switches can cause such RFI [69]. All devices using RF can be vulnerable to RFI. Unintentional interference can be either internal or external. Figure 5 shows some examples of unintentional RF interference.

Using spectrum analyzers RFI can be recognized by monitoring values such as bit error rate and noise. When RF interference is present, it affects the antenna temperature, causing disruption of geophysical parameters [24].

Mitigation techniques for detecting RF interference consist of time and frequency domain based methods and spectrograms. Time and frequency domain based methods remove power samples which are higher than expected values. Spectrograms use series of data to receive good resolution in both time and frequency.

Increasingly accessible low cost hardware and recent technological developments in cognitive networking and software-defined radio (SDR) have made many applications dependent on wireless networks. This gives adversaries the opportunity to inflict harm or damage to systems relying on wireless communication networks through jamming attacks (also known as intentional RFI) [97]. This type of attack causes a DoS problem. DoS attacks can result in slower download times, greatly reducing the number of active voice users or a significant increase in network latency [29]. Due to the simplicity of jamming techniques, there exists a variety of inexpensive jamming technologies available to jammers, making it difficult to overcome these types of attacks [35]. For securing wireless communication systems and guaranteeing quality of service (QoS), a robust RFI detection method needs to be applied for producing an effective mitigation process [35]. Moreover, determining the modulation type of the signal of interest combined with RFI precisely is essential. For enabling demodulation processes at the receiver side, automatic modulation classification (AMC) rises as an important procedure in communication networks [22].

4.4.1 Reactive, Non-adaptive Traditional Approaches

Out-of-band emission by the meaconers, jammers, spoofers, close transmitters and harmonics from other intefereing sources are the most common reasons for RFI.

RFI is prevailing for these emitters in global navigation satellite systems (GNSS), [12], microwave radiometry [36], and radio astronomy [47, 93]. In addition, it exists for imperfect sensing spectrum in cognitive radio system [31]. Consequently, RFI detection and excision bring a necessity for development of statistical [81], transformed domain-based [23], spectral [36], temporal [68], spectral-temporal [12], and spatial filtering-based [47, 93] techniques.

The kurtosis method [81] detects non-Gaussian distributed interference signals efficiently and considers those signals as RFI; however, the kurtosis method is not able to detect Gaussian or approximately Gaussian distributed interference signals. The paper [68] proposes asynchronous pulse deletion method which performs well specifically for impulse interference. Nevertheless, the performance of the asynchronous pulse deletion technique becomes quite susceptible to chosen detection parameters. Power detection techniques like crossfrequency technique [36] works at frequency domains. They are able to mitigate little RFI successfully and efficiently for many channels. Spatial methods detect RFI by estimating its subspace. In the sequel, these techniques project the estimated RFI subspace for eliminating RFI [36].

For microwave radiometry applications, the researchers have proposed spectral detection and excision techniques like the mitigation technique in [16] and the crossfrequency blanking [36]. These techniques commonly apply reconstructed interference and fast Fourier transforms (FFT) for mitigating RFI. However, crossfrequency blanking needs detection thresholds because they may decrease performance, which implies the mitigation technique becomes unsuitable for wideband RFI mitigation if the thresholds are set incorrectly. As an example of temporal algorithms, asynchronous pulse blanking is a commonly used technique. It blanks the part at which signal amplitude becomes larger than the threshold with respect to noise. Nevertheless, asynchronous pulse blanking has a performance decrease because of exploited heuristic threshold.

An interfering signal emerges for a short duration in many cases and demonstrates a variable frequency behaviour [23]. A time-frequency representation like Gabor expansion or a spectrogram facilitates the identification and removing of RFI in such cases [12]. In [23], the RFI is estimated in time-frequency domain and subtracted from the signal. In a similar way, transformed domain-based techniques using the Karhunen-Loève (KL) and wavelet transform, bordered autocorrelation techniques were presented in a detailed way in [23] and [59], respectively. Nevertheless, techniques presented in [23] have high computational complexity; the techniques presented in [59] cannot detect wideband signals unambiguously.

Despite their advantages, transformed domain-based and time-frequency techniques are highly complex computationally. Hence, one may apply statistical techniques like kurtosis detection [81] by assuming non-Gaussian RFI.

Besides, the paper [8] proposes statistical techniques assuming unknown statistics for RFI. The work [8] applies generalized likelihood ratio test and Neyman-Pearson detection theory to derive a novel GNSS detection technique. However, because of their non-linear operations, the statistical detection and excision techniques are computationally complex.

The previously mentioned highly-complex techniques are also susceptible to RFI misdetection. Finally, signal processing practitioners may opt for spatial filtering methods like cross subspace projection (CSP) [47] and subspace projection (SP) [93]. SP relies on eigenvalue decompositions of space-time autocorrelation matrices whereas CSP relies on singular value decomposition (SVD) of the space-time crosscorrelation matrix. Particularly for radio astronomy applications, they are state-of-the-art methods in order to take away the RFI which has more stable interferer emission. In addition, the paper [38] proposes oblique projection beamforming technique to cyclostationary RFI mitigation. Recently, the scenarios with correlated noise, relatively rapid interference motion and little interference-to-noise ratio have been addressed by the polynomial-augmented subspace projection technique [54].

4.5 *Anti-spoofing Solutions*

Spoofing can be described as making GPS receivers calculate false positions.

For spoofing receivers, an adversary requires recreating signals from multiple satellites faithfully. In the sequel, it sends those "spoofing" signals for capturing local GPS receivers. If targeted GPS receivers cannot distinguish spoofed signals from real satellite signals, spoofing can mislead target receivers such that they appear to be at different locations.

Two application areas of anti-spoofing are facial authentication (FA) systems and global navigation satellite systems.

4.5.1 Facial Authentication Systems

FA systems have recently been used in daily applications such as individual identification, online payment and access control. Because of its convenience and precision, FA is regarded as a more promising authentication approach different from traditional ones, like token, fingerprint, and PIN code [101].

Being mostly camera-based, brings some severe drawbacks to the existing FA systems, which include security problems and privacy leakage risks. Nevertheless, many existing FA systems use RGB cameras for collecting facial features of users, which reveals complete visual facial information (VFI). This contributes to exacerbating privacy concerns. Nonetheless, by deriving geometric features from VFI, this approach performs authentication. By reproducing the features, an attacker can make spoofing attacks easily provided that a user VFI is leaked.

Even though the research literature has introduced more geometry information such as depth information of faces [51] for enhancing security, a camera-based FA system is yet susceptible to spoofing attack since it can capture information remotely. Examples include using infrared dot projectors or depth cameras [99]. Therefore, attackers can manipulate 3D-printed VFI-captured-based masks for deceiving FA systems [25].

To overcome the disadvantages, bio-material-based FA can be used since it is more resilient against spoofing attack. At the same time, the recent developments in wireless sensing imply the RF signals are susceptible to materials that RF signals encounter during propagation [101]. Therefore, authenticity of users can be determined by verifying VFI captured in RF signal. Moreover, the scholars demonstrate that an array of commercial-off-the-shelf (COTS) RFID tags are able to work properly as sensitive and cheap sensors for supporting fine-grained wireless sensing [95]. As RFID systems have the fine-grained sensing and material-sensitive capabilities, an anti-spoofing FA system is attempted to be developed with COTS RFID devices.

4.5.2 Global Navigation Satellite Systems (GNSS)

GNSS plays an important role in location, exploration, and communications [53]. Unfortunately, the vulnerability of GNSS signals to spoof interference has long been known. By broadcasting counterfeit GNSS signal, a spoofer tries to deceive GNSS receivers. Counterfeit signals have similar structure as genuine signals, which may result in false decisions to receivers. Thus, one of the crucial tasks of GNSS is to distinguish spoofing signals from real signals.

For individual transmitter identification, RF fingerprinting has been used widely. The features of RF fingerprinting mainly include steady-state and transient-state features broadly. During transmitter's power off and on, a transient signal occurs. This signal is also very short. It is very hard to capture transient signals so transient-state features have limited applications. The durations of steady-state signals are longer than those of transient signals. They are easier to be captured than the transient ones. Therefore, it is more practical for studying.

Some of the present steady-state feature extraction approaches include higher order spectra analysis, fractal theory, and time-frequency transform. Higher order spectra is suitable for classification, because higher order spectra has some great characteristics like phase information of signals, keeping the amplitude and handling Gaussian noise. However, this technique requires assumption of stationarity of the signal. Nevertheless, GNSS signals are generally non-stationary and non-linear; therefore, some more efficient methods need to be found [90].

The work [32] presents Wigner bispectrum (WB), bispectrum-based time-varying higher order spectra, and Wigner higher order spectrum [27] can be described as extensions of WB to higher order spectra domain. Nevertheless, WB's direct utilization needs 3D matching template which makes it computationally complex. Diagonal sliced Wigner bispectrum (SWB) is one of the main dimension reduction methods for reducing computational complexity. By getting a 2D WB slice, SWB is obtained. It can just use a little portion of all WB information. The paper [90] proposes axial integrated Wigner bispectrum (AIWB) as a novel dimension reduction method. AIWB integrates WB along parallel direction with a frequency axis on the bi-frequency plane. The AIWB both reduces WB's dimension and uses almost all WB information. Numerical results demonstrate that this proposed method has superior performance compared with bispectrum and SWB.

4.5.3 High-End and Low-End Receivers

ZigBee devices which use IEEE 802.15.4 standards of WPAN have commonly been adopted in numerous applications such as industrial and building control, healthcare, and security systems [71]. Their rising popularities in high-value and sensitive fields bring vulnerability and security concerns. Advanced encryption standard (AES) provides ZigBee security for Application, Network and MAC layers. Nonetheless, several techniques have been proposed for exploiting vulnerabilities during key-exchange process [21]. Moreover, it is shown that via side-channel analysis (SCA) techniques, it can passively monitor power consumed by authorized wireless sensor nodes for determining secret encryption key [78]. ZigBee alliance overcomes the security issues by adopting AES in a more resilient mode against an SCA attack, either or cipher block chaining (AES-CCM) or counter (AES-CTR) mode. Nevertheless, AES-CTR mode is exhibited to be susceptible to SCA eavesdropping attack theoretically and practically in [46] where full AES-CTR key has been recovered in a successful way. Hence, in ZigBee devices, key recovery becomes totally possible, which enables attackers to insert rogue devices into present networks.

5 Machine and Deep Learning-Based RF Security for Cyber-Physical Systems

In this subsection, we present machine learning (ML) and deep learning (DL) based security solutions for RF security problems in RF fingerprinting, interference detection, anti-spoofing and antijamming fields.

5.1 Machine Learning Based Approaches for RF Fingerprinting

This subsection presents ML based RF Fingerprinting techniques for RF Security.

ML and DL techniques are known to be successful especially for extracting specific latent features contained in the data. Traditional RF fingerprinting methods rely on built-in measures or some protocol-specific techniques [48]. Deep learning can be used for detecting patterns and it can outperform detecting feature better than handcrafted features.

In the literature, there are several works that apply machine learning for detecting RF fingerprints. In [40], RF fingerprinting is performed by extracting features such as constellation feature, and phase noise spectrum in the transmitted waveform.

In [98], long short term memory (LSTM)-based recurrent neural network (RNN) is developed and used for detecting hardware properties. LSTM can learn from past values and retain information. This way, the knowledge of the past transmitter information can be used. Deep learning methods do not require human intervention for deciding features. In that research, single-carrier quadrature phase shift keying (QPSK) modulation is considered. Gaussian noise is added on that signal for simulating decay of signal-to-noise-ratio (SNR). Without filter or recover, I/Q data is collected at analog-to-digital converter (ADC) output.

Different architectures of convolutional neural network (CNN) is proposed in [48] to detect specific transmitters. With the use of spectrum analyzer, they have captured IQ samples that contains Wi-Fi signals of 5117 devices with 166 transmissions for each one and ADS-B samples of $5,000$ devices with 76 transmissions for each one, respectively. As a neural network architecture, the CNN based on AlexNet and ResNet is utilized. Both architectures performance has been investigate for varies SNR regimes. Morevore, channel effect is studied since the dataset contains both Wi-Fi and ADS-B signals.

Another research that investigates Wi-Fi and ADS-B protocols is proposed by Cekic et al. [13]. Differing from the above mentioned work, this research investigates use of complex valued neural networks for RF fingerprinting. The focused features are both IQ imbalance, and differential and power amplifier non-linearity. The effect of the multi path channels is also investigated. The results are compared for both with augmentation and without augmentation. This is performed by adding additive Gaussian white noise (AWGN).

Belgiovine et al. [82] propose raw IQ samples based studies for static channels. As such, the for static channels no channel prediction are required however, for the channels do not need to be predicted. For varying channels, the model is trained with complex demodulated symbols to eliminate the channel effect. Their neural architecture is based on AlexNet. The bit-error-rate (BER) value of IQ imbalance and BER vs. DC offset level for different SNRs are investigated. The obtained accuracy is 99%. It can be understood that the proposed model is capable of detecting devices using impairments for both varying channel and static channel conditions.

Ozturk et al. [70] focuses on both RF time series images and the spectrograms of radio signals of drone controllers by utilizing CNN network. In time series it is mentioned that when the SNR decreases, the model does not perform well because transient signal is lost in the noise. However, with spectrograms, it is possible to focus on frequency. Time series images are used for training the neural network. Images are obtained by plotting the RF arrays. Since the experiments are conducted in laboratory, noise is added to the data to obtain more realistic simulations.

Mohanti et al. [63], proposed RF fingerprinting approach to distinguish authorized unmanned aerial vehicle (UAV). Transmitted IQ samples are collected from each UAV for detecting specific fingerprint using CNN. In experiments both ground receivers and mounted receivers on UAVs are used. In this work, the focus is adding amplitude changes in real/imaginary parts, to obtain only IQ imbalance. Constellation points are used to detect these impairments. As a consequence, an ML classifier trained in one channel environment cannot perform well if the wireless channel changes. To overcome this challenge, processing block is designed at the transmitter side for adjusting the IQ samples before transmission. The dataset consists of samples from seven UAVs which are of the same make and model rather than different models as used in many research works. The used neural network architecture in this research, is a one dimensional version of the very deep Convolutional (VGG) network [88]. The authors were able to achieve accuracy of 98% for UAV detection based on their CNN model that consists of seven convolution layers with maxpooling layers and three fully dense layers.

5.2 Proactive, Adaptive Machine Learning Based Approaches for RF Interference Detection

This subsection presents ML based RF interference detection techniques for RF Security. With fast developments in the artificial intelligence (AI) technology, more and more research studies have deployed DL to RFI field and the classification of modulation. This subsection explains some of these algorithms which apply ML techniques for RFI detection.

Mosiane et al. [64] apply conventional supervised ML algorithms like decision tree forests, nearest K-neighbors, and naive Bayesian classification for classifying the signals of RFI. Before applying the classifier, the time consuming pre-processing of feature extraction process is required. Besides, it cannot guarantee to extract deep data characteristics well. Therefore, A keret et al. [2] propose the usage of U-Net, a specific kind of CNN to detect and mitigate RF interference. In that paper, CNN architecture is trained by simulated data. As RFI demonstrates a non-typical behaviour compared with normal signal, Ghanney et al. [33] study RF interference detection as anomaly detection problem. Chalapath et al. [14] apply DL to detect anomalies while many advanced unsupervised DL techniques have been proposed. Some of those techniques can be listed as generative adversarial networks (GANs) [87], LSTM [60] and autoencoders (AEs) [5, 79].

For detecting clutter and interference in high frequency surface wave radars, Zhang et al. [106] propose faster region-based convolution neural network. As input for this network, it uses range-Doppler spectrum image. Consequently, this proposed technique has a decent detection performance and high classification accuracy [106].

Yang et al. [103] develop a CNN-based algorithm RFI-Net for detecting interference at the 500-m aperture spherical radio telescope. This technique outperforms CNN-based U-Net model, sum-threshold, and k-nearest neighbors. Gecgel et al. [28] use deep RNNs and deep convolutional neural networks to detect jamming attacks. This paper analyzes reference signal jamming and classical wide-band barrage jamming. Numerical results in this paper demonstrate that under realistic test environment, classification accuracy reaches up to 86.1%.

Ramjee et al. [75] have proposed deep residual network, LSTM neural network and convolutional long short-term deep deural network for recognizing 10 different modulation types. It is numerically demonstrated that at high SNRs, the techniques have increased classification accuracy by up to 90%. By minimising training dataset size, Ramjee et al. [75] have also deployed principal component analysis (PCA) for optimizing classification process. Jiang et al. [49] present a technique combining pretrained inception-ResNetV2 with the transfer learning for recognizing 3 kinds of modulation at 4 dB SNR, which are binary phase shift keying (BPSK), QPSK, and 8PSK. Numerical result in [49] demonstrates that classifications accuracy for recognizing BPSK, QPSK, 8PSK are 100%, 99.66%, 96.33%, respectively.

Karra et al. [50] present a robust hierarchical deep neural network (DNN) architecture performing a hierarchical classification for estimating modulation order, modulation class, and data type (digital or analog modulation). As input of CNN, authors utilize snapshot of spectrogram derived from in-phase, quadratic component of baseband signals and for most modulation schemes to achieve 90% at high SNR. Yang et al. [102] present efficient method which uses RNN and CNN for classifying 6 modulation types under Rayleigh fading and additive white Gaussian noise (AWGN). Experimental results demonstrate that classification precision for CNN becomes nearly 100% in AWGN channel [102].

Minimum classification accuracy still approaches 84% even in Rayleigh channel, while maximum value approaches to 96%. Shi et al. [84] present a robust CNN-based technique that can classify BPSK, QPSK, 8PSK, and 16QAM in an orthogonal frequency division multiplexing (OFDM) system precisely under phase offset presence. Zhang et al. [104] developed scheme based on LSTM and CNN for solving the AMC problem.

Moreover, suggested fusion model-based classifiers in parallel and serial modes become quite useful for improving classification accuracy for SNRs varying from 0 dB to 20 dB [104]. Serial fusion mode achieves better results than others as shown in [104]. Ujan et al. [91] tackle the RFI problem in a different way. Besides specifying the type of received signal, it also determines various schemes of digital modulation in which the jamming signals exist in the DVB-S2 standard-based real-time digital video broadcasting by using transfer learning.

A hierarchical classification technique is proposed for classifying RFI and AMC by taking advantages from transfer learning technology which uses pretrained CNN like GoogleNet, ResNet18, VGG16, AlexNet for feature learning to which a full-connected classifier follows. The paper analyzes these pretrained CNNs comparatively with respect to the accuracy in context of consumed training time and transfer learning. The visual representations of received signal have been generated as the input data in time-frequency domain. This is known as scalogram which is magnitude squared of wavelet transforms.

Ghanney et al. [33] consider RFI as anomalous distorting parasite signal since it has a damaging influence on wireless communications. Consequently, the RFI mitigation becomes quite essential for avoiding this impact. RFI detection and localization can be considered as initial steps for RFI mitigation process. This work proposes the following approaches for detecting and localizing RFI by using unsupervised and supervised DL techniques. Firstly, it studies a CNN-based object detection algorithm as the supervised approach. This technique is based on you only look once v3 (YOLO-v3) trained on real-world data which is contaminated by multiple RFI source. Secondly, as an unsupervised approach, it proposes usage of convolutional auto encoder (CAE). The numerical results demonstrate that RFI detection by YOLO-v3 is faster and it achieves 94% accurate detection rate and demonstrates that its average precision can achieve 89% accuracy. Average precision of the CAE can achieve 78% and performs better than YOLO-v3 in certain cases.

5.3 Machine Learning Based Anti-spoofing Approaches for RF Security

This subsection presents ML based anti-spoofing techniques for RF Security.

Xu et al. [101] propose RFace as a novel anti-spoofing privacy-preserving authentication system. In this strategy, an RFID tag array is used for measuring phase and RSS of RF signal in order to derive bio-material and 3D facial geometry features to make spoofing attacks resilient. A well-trained Support vector machine (SVM) classifier uses the extracted features for conducting authentication. In addition, RFace achieves resiliency against spoofing attacks because it is difficult for attackers to capture the extracted features.

RF fingerprinting provides catchy countermeasure in physical layer against spoofing attacks and rogue device identification. Physical differences between devices manifest themselves into measurable differences in their sent signals. RF-fingerprinting applies ML techniques for exploiting the differences between wireless devices and identifying them reliably [77]. RF fingerprinting is an efficient solution in general since it utilizes stochastic physical variations between devices to identify them, which can be considered as prohibitively for duplication of an attacker.

Most RF fingerprinting techniques require using high-speed oscilloscopes, or expensive, large receivers like Agilent E3238S [71]. Recent research in RF fingerprinting has been successful via usage of low cost universal software radio peripheral (USRP) receivers [77]. Nevertheless, no study has compared low and high cost receivers for performance tradeoffs. Patel et al. [71] compare the differences in performances of RF-fingerprinting between low-cost USRP-based receivers and high-cost receivers. Testing is conducted by using six devices which have identical model type from the same manufacturer. The uniformity in the model demonstrates hardest case for device identification, and performance can be improved just for cases with different device models. Random forest classifier is used for investigating the features of amplitude, frequency and phase of the signals and their respective contribution in classification. The correct classification rate (%C) characterize its performance to identify devices. It also identifies rogue device scenarios for network intrusion detection.

5.4 Machine Learning Based Antijamming Approaches for RF Security

This subsection presents ML based antijamming techniques for RF Security.

For detecting jamming attacks, the literature include ML methods based on classifiers like SVM and artificial neural network (ANN) with different features. In order to illustrate, Punal et al. [73] present an ANN-based technique for wideband spectrum sensing and cyclic spectral analysis. Based on modulation and signal quality, this technique discriminates jamming signals from narrowband ones. Grover et al. propose a ML based jamming detection system via expectation maximization, SVM, and adaptive boosting techniques. Maximum inactive time, busy channel ratio, packet delivery ratio, and noise have been used for detecting jamming attack. Many present methods need more resource and serve just as stop-gap ultimately. They can detect link states; however, these methods cannot detect the source of service outage. Moreover, these methods have larger false alarm rates. They need accurate algorithms for training and testing the classification models. Features selection and learning curves are often neglected while these are some of the most important aspects in designing detection techniques with machine learning. Thus, there is a great need for efficient and fast detection techniques able to detect jamming attacks more accurately.

Arjoune et al. [7] propose a ML-based technique for detecting transmission link state between receiver and transmitter for checking whether the link is attacked. ML-driven techniques can achieve high detection accuracy provided that they consider the following stages: select suitable input features, collect and build a large dataset, and use correct methods for training, validating, and testing the model. For detecting jamming attacks, the used features and parameters include clear channel assessment, bad packet ratio, packet delivery ratio and received signal strength. The paper studies methods for assessing communication link status and selecting appropriate features. A large dataset is built for training, validating, and testing ML models. Cross-validation techniques, dataset normalization and randomization were performed to avoid underfitting.

Ak et al. [1] study antijamming performance of cognitive radar under partially observable Markov decision process (POMDP) models. To begin with, an explicit expression is obtained for jammer dynamics uncertainty, which enables us to discover new insights into the probability for the radar to be jammed beyond conventional SNR-based analysis. By applying two frequency hopping strategies proposed for reinforcement learning (RL) frameworks, the performance metric is used in LSTM network and deep Q-network (DQN) under various uncertainty values. Finally, for both network architectures, a softmax operator replaces the target network require-

84. Shi J, Hong S, Cai C, Wang Y, Huang H, Gui G (2020) Deep learning-based automatic modulation recognition method in the presence of phase offset. IEEE Access 8:42841–42847
85. Ski J, Wan J, Yan H, Suo H (2011) A survey on cyber physical systems. In: 2011 international conference on wireless communications and signal processing (WCSP), pp 1–6. https://doi.org/10.1109/WCSP.2011.6096958
86. Saucedo-Martínez JA, Pérez-Lara M, Marmolejo-Saucedo JA, Salais-Fierro TE, Vasant P (2018) Industry 4.0 framework for management and operations: a review. J Ambient Intell Hum Comput 9:789–801. https://doi.org/10.1007/s12652-017-0533-1
87. Schlegl T, Seebock P, Waldstein SM, Schmidt-Erfurth U, Langs G (2017) Unsupervised anomaly detection with generative adversarial networks to guide marker discovery. In: Proceedings of IEEE international conference on information processing in medical imaging. Springer, Berlin, pp 146–157
88. Simonyan K, Zisserman A (2015) Very deep convolutional networks for large-scale image recognition. In: International conference on learning representations
89. Sun X, Ng K, Ding Z, Xu Y, Zhong Z (2019) Physical layer security in UAV systems: challenges and opportunities. IEEE Wirel Commun 26(5):40–47. https://doi.org/10.1109/MWC.001.1900028
90. Sun M, Zhang L, Bao J, Yan Y (2017) RF fingerprint extraction for GNSS anti-spoofing using axial integrated Wigner bispectrum. J Inform Secur Appl 35:51–54
91. Ujan S, Navidi N Jr, Landry R (2020) An efficient radio frequency Interference (RFI) recognition and characterization using end-to-end transfer learning. Appl Sci 10:6885
92. Utne IB, Hassel H, Johansson J (2012) A brief overview of some methods and approaches for investigating interdependencies in critical infrastructures. In: Hokstad P, Utne I, Vatn J (eds) Risk and interdependencies in critical infrastructures. Springer series in reliability engineering. Springer, London. https://doi.org/10.1007/978-1-4471-4661-2-1
93. van der Tol S, van der Veen A-J (2005) Performance analysis of spatial filtering of RF interference in radio astronomy'. IEEE Trans Signal Process. 53(3):896–910
94. Wang, B, Wang Y, Chen R (2019) A practical authentication framework for VANETs. Secur Commun Netw 2019 (Article ID 4752612)
95. Wang C, Liu J, Chen Y, Liu H, Xie L, Wang W, He B, Lu S (2018) Multi-touch in the air: device-free finger tracking and gesture recognition via COTS RFID. In: IEEE international conference on computer communications, INFOCOM
96. Wang W, Sun Z, Piao S, Zhu B, Ren K (2016) Wireless physicallayer identification: modeling and validation. IEEE Trans Inf Forensics Secur 11(9):2091–2106
97. Weerasinghe S, Alpcan T, Erfani SM, Leckie C, Pourbeik P, Riddle J (2018) Deep learning based game-theoretical approach to evade jamming attacks. In: Bushnell L, Poovendran R, Basar T (eds) Decision and game theory for security. GameSec 2018. Lecture notes in computer science, vol 11199. Springer, Cham. https://doi.org/10.1007/978-3-030-01554-1_22
98. Wu Q, Feres C, Kuzmenko D, Zhi D, Yu Z, Liu X, Liu X (2018) Deep learning based RF fingerprinting for device identification and wireless security. Electron Lett 54:1405–1407
99. Wyatt A (2018) What exactly is the dot projector? Why it is used in iphone x? https://www.thebestintech.com/what-is-dot-projector/
100. Xu Q, Zheng R, Saad W, Han Z (2015) Device fingerprinting in wireless networks: challenges and opportunities. https://arxiv.org/pdf/1501.01367.pdf Cited 28 Dec 2021
101. Xu W, Liu J, Zhang S, Zheng Y, Lin F, Han J, Xiao F, Ren K (2021) RFace: anti-spoofing facial authentication using COTS RFID. In: IEEE international conference on computer Communications (INFOCOM 2021), pp 1–10
102. Yang C, He Z, Peng Y, Wang Y, Yang J (2019) Deep learning aided method for automatic modulation recognition. IEEE Access 7:109063–109068
103. Yang Z, Yu C, Xiao J, Zhang B (2020) Deep residual detection of radio frequency interference for FAST. Mon Not R Astron Soc 492:1421–1431
104. Zhang D, Ding W, Zhang B, Xie C, Li H, Liu C, Han J (2018) Automatic modulation classification based on deep learning for unmanned aerial vehicles. Sensors 18:924

105. Zhang J, Woods R, Sandell M, Valkama M, Marshall A, Cavallaro J () Radio frequency fingerprint identification for narrowband systems, modelling and classification. IEEE Trans Inf Forensics Secur 16:3974–3987
106. Zhang L, You W, Wu Q, Qi S, Ji Y (2018) Deep learning-based automatic clutter/interference detection for HFSWR. Remote Sens 10:1517
107. Zhuo F, Huang Y, Chen J (2017) Radio frequency fingerprint extraction of radio Emitter based on I/Q imbalance. Procedia Comput Sci 107:472–477
108. Zio E (2016) Critical infrastructures vulnerability and risk analysis. Eur J Secur Res 1:97–114

Attack Detection by Using Deep Learning for Cyber-Physical System

Saeid Jamshidi, Amin Nikanjam, Mohammad Adnan Hamdaqa, and Foutse Khomh

Abstract With a cyber-physical system (CPS), physical components like industries are handled with an automated system. With the booming of cyber-attacks, detecting these attacks remains challenging. In order to protect the system from being hacked, we need to have CPS security measures implemented. Machine Learning (ML) has an important role to play in the detection of security attacks, which is the first step to protecting the CPS system. Cutting edge Deep Learning (DL) techniques have widely been applied to various domains like image processing and speech recognition. As part of a review of detecting cyber-attacks in CPSs, this chapter outlines the roles of DL and Deep Reinforcement Learning (DRL). Also, we present state-of-the-art solutions without sacrificing technical details. Additionally, we describe common datasets used for DL in CPSs. Finally, we express research opportunities and challenges in the CPSs with respect to DL.

Keywords Attack detection · Cyber-physical system · Deep learning · Dataset · Cyberattack · Attack type · Machine learning · Vulnerabilities · Threats · Adaptability · Auto encoder · Data acquisition

S. Jamshidi (✉) · A. Nikanjam · F. Khomh
SWAT Laboratory, Polytechnique Montréal, Montréal, Canada
e-mail: jamshidi.saeid@polymtl.ca

A. Nikanjam
e-mail: amin.nikanjam@polymtl.ca

F. Khomh
e-mail: foutse.khomh@polymtl.ca

M. A. Hamdaqa
SæT Laboratory, Polytechnique Montréal, Montréal, Canada
e-mail: mohammad-adnan.hamdaqa@polymtl.ca

I. Traore et al. (eds.), *Artificial Intelligence for Cyber-Physical Systems Hardening*, Engineering Cyber-Physical Systems and Critical Infrastructures 2,
https://doi.org/10.1007/978-3-031-16237-4_7

1 Introduction

A typical CPS consists of a computer, controller, and physical system. CPSs operate using algorithms that are based on computers. The physical and non-physical components of CPSs are interdependent, and their interdependence is essential for the proper functioning of the system. A sensor data analysis may be used to detect anomalies in CPS data so that abnormal behavior may be detected. As an example, it could be used for detecting security breaches, problems, etc. Cyber security involves detecting various communication attacks that lead to malfunctions in physical systems, thereby compromising the objectives that the physical system was designed to accomplish.

Also, the CPS market is expected to grow at a rate of 9.7% per year by 2025, which will result in annual revenue of $9563M according to a report by [1]. As part of the aforementioned, hackers are also exploiting the COVID-19 pandemic and shifting from onsite to remote workstations, increasing the number of cyber-attacks by 29% [1]. The number of ransomware attacks has risen at least 93% in 2021, possibly due to improved attack techniques such as Triple Extortion [2]. Cyberattacks related to the Internet of Things (IoT) are expected to double between 2015 and 2025, contributing to the growth of this risky industry in 2021. Cyberattacks have become more sophisticated in response to the increasing prevalence of automated attack tools [1, 2], and professional hacking groups are heavily using these tools. Table 1 shows some of the real attacks against the physical systems. Cyber solutions must be capable of addressing well-scoped problems. In order to work together with the new technology, the tools and architecture of the current system must be compatible, and also it should be possible to evaluate system performance without any difficulty. Machine Learning (ML) is becoming an increasingly important concept for cyber security. A ML technique can be applied in many different ways, including regression analysis in cyber security for fraud detection, classification approach in spam filters, clustering used for forensic analysis, and dimensionality reduction in facial recognition.

It is no secret that Deep Learning (DL) has grown in popularity in recent years as it has led to significant improvements in many applications domains including security-related applications in critical infrastructures [4], like detecting attacks and unexpected errors in critical infrastructures [5]. A Deep Neural Network (DNN) with several types of layers presents higher-level representations containing amplified features (points of difference between samples) that can effectively discriminate certain samples from irrelevant ones, while at the same time suppressing unnecessary samples. Through the application of DL models, researchers have been able to improve the performance and accuracy in many different areas of Artificial Intelligence (AI). Examples include speech recognition, object detection, natural language processing, and pattern recognition [5]. There are several advantages of using DL [6, 7] over traditional ML, for example in speech recognition, statistical arbitrage and medical diagnosis, DL models usually demonstrate superior results, provided that the data is sufficient [8, 9]. The past few years have seen the development of some DL models to detect cyberattacks targeting critical infrastructures, and alert intrusion

Table 1 Real CPS attacks [3]

Country	Target	Attack nature	Type	Date	Motives
United States of America	Ohio Nuke Plant Network	Slammer Worm	Malware-Dos	January 25, 2003	Criminal
	Taum Sauk Hydroelectric Power Station Failure	Sensors Failure	Accidents	December 14, 2005	N/A
	Georgia Nuclear Power Plant Shutdown	Installed Software Update	Undefined Software	March 7,2008	Unclear
	US Electricity Grid	Reconnaissance	Undefined Software programs	April 8, 2009	Political
	Springfield Pumping Station	Backdoor	Unauthorized Access	November 8, 2011	Criminal
	Georgia Water Treatment Plant	Physical Breach	Unauthorized Access	April 26, 2013	Criminal
Iran	Iranian nuclear facilities power plant and other industries Iranian infrastructure (nuclear, oil)	Stuxnet Stuxnet-2 DDoS	Worm Worm Disruptive	November 2007 December 25, 2017 October 03, 2012	Political Political Political
	Communication companies Iranian key oil facilities	Computer virus	Malware	April 23, 2012	Political
Saudi Arabia	Saudi infrastructure in the energy industry	Shamoon-1	Malware	August 15–17,2012	Religio-Political
	Saudi government computers and targets	Shamoon-1	Malware	November 17,2016	Religio-Political
	Tasnee and other petrochemical firms, National Industrialization Company, Sadara Chemical Company	Shamoon-1	Malware	January 23,2017	Religio-Political

(continued)

Table 1 (continued)

Country	Target	Attack nature	Type	Date	Motives
Qatar	Qatar's RasGas	Shamoon	Malware	August 30, 2012	Political
United Arab Emirates	UAE Energy Sector	Trojan Laziok	Malware	January–February 2015	Political
Australia	Maroochy Water Breach	Remote Access	Unauthorized Access	March, 2000	Criminal
Canada	Telvent Company	Security Breach	Exploited Vulnerability	September 10, 2012	Criminal
Ukraine	Ukrainian Power-grids	BlackEnergy Malware	DDoS	December 23, 2015	Political
	Ukrainian Electricity Firms	Petya	Ransomware	June 27, 2017	Political

detection, malware detection, access control, anomaly detection, and classifications [5].

On the other hand, the data gathered from CPSs are usually multidimensional and DL models are specifically designed to handle such data that has several dimensions. Another characteristic of CPSs is that they continually grow in data volumes, they are subject to concept drift, and new threats constantly attempt to compromise their operations. For this reason, security solutions for CPS data-driven devices must be developed that are capable of adapting and expanding to changes in data. They should be able to continuously discover new threats and vulnerabilities within. Nonetheless, there is a challenge in generalizing the adaptability concept in the context of security-based applications built for CPSs. It is almost impossible to use a developed model for a particular scenario in another one, even if these are part of the same process. The performance of DL models in a real-world situation depends mainly on its generalization capability, which is decided by how the model handles the data which it is not familiar with, i.e., how well it can deal with new data without any knowledge of its priors [10].

The purpose of the chapter is to review the advances in detecting cyber-attacks made possible by DL-driven solutions in the CPS domain. Readers are provided with an overview of utilizing a DL-driven methodology consisting of multiple stages to help them grasp the process more quickly and develop the skills they need. A multi-step methodology is used to evaluate DL performance by considering the full cycle of DL scenarios through performance evaluation. Researchers, practitioners, and students will benefit from this chapter's focus on building cyber security applications using DL-based methods.

The rest of the chapter is organized as follows: Sect. 2 gives an overview of DL in CPSs. Section 3 discusses DL techniques. Section 4 discusses the state-of-art DL techniques that have been successfully used in the field of cyber-security/CPSs. Section 5 discusses RL and DRL in CPSs. Section 6 introduces data acquisition in CPSs. Section 7 discusses the challenges to attack detection in CPSs. Section 8 outlines robust attacks detection. Finally, Sect. 9 makes concluding remarks.

2 CPSs and DLs

CPS is one of the newest technologies that provides the integration of computer systems, physical systems, and control systems together. A part of the CPS is the automation of all the industrial processes; this includes the manufacturing, monitoring, and control processes. Due to the fact that the system involves three different domains of optimization, namely computational, physical, and control, these types of systems are naturally complex in nature and therefore cannot be optimized conventionally. In order to design and optimize such complex systems, ML provides effective ways to model the behavior of such systems. Figure 1 illustrates the various applications of ML to CPSs, including:

- Detection of anomalies,
- Security and cybercrime,
- Identifying faults,
- Maintaining predictively,
- Optimization of the process,
- Analysis of QoS,
- Allocating resources.

ML algorithms, as a branch of AI, have been used to enhance the effectiveness of many systems [11]. The reference of data and AI as "new oil" and "new electricity", respectively, underscores their impact in the present world [12]. Furthermore, AI

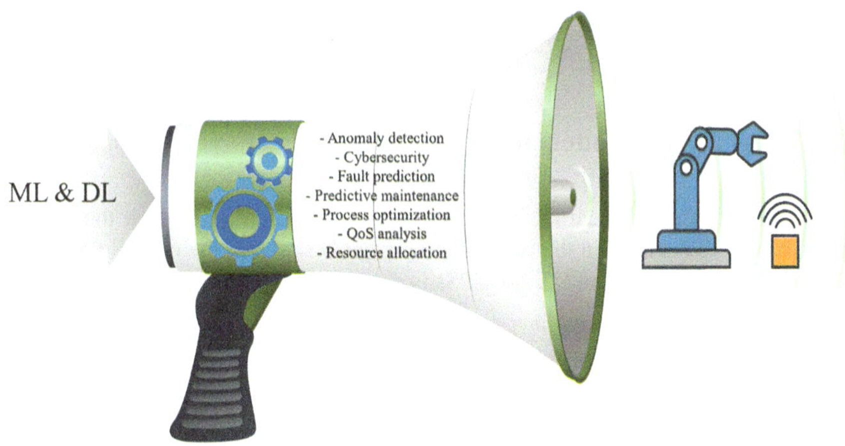

Fig. 1 Application of ML in CPS

along with other emerging technologies like IoT, and CPS are the major technologies pushing for the fourth industrial revolution [13]. As a result of these developments, a key area of research interest is securing systems from adversarial attacks using AI.

Some of the early applications of ML for cybersecurity is in Intrusion Detection System (IDS). Research in this area included malware and anomaly detection in information and communication systems. With the success recorded, ML was also used to achieve cybersecurity in IoT systems [14, 15]. Furthermore, the combination of DL and Reinforcement Learning (RL) have contributed significantly to solve problems that posed a challenge to shallow algorithms and the more familiar supervised/unsupervised algorithms. Factors that support the use of DL in CPS include the high-dimensional data generated and the continual growth of data [16].

Moreover, DL techniques have gained a high focus in data science as they have enhanced performance in many applications. DL algorithms consist of hierarchical architectures with multiple layers in which lower-level features are transferred to higher-level ones. They have the capability of extracting relevant features from the underlying data. Moreover, the notion of DL combined with RL is also one of the best tools we have at our disposal to cope with unstructured environments; they can learn from large amounts of data or discover patterns [17]. DL has successfully been applied in CPSs for security applications such as intrusion detection, malware detection vulnerability identification in CPSs [5]. Due to the fact that CPSs produce large volumes of data generated by numerous sensors, DL is suitable for this setting. Since DL methods rely primarily on neural networks as the basis of computing, the term "DNN" is often used to refer to DL models.

Recently, researchers have combined DL and RL to arrive at Deep Reinforcement Learning (DRL); a development that has resulted in a tremendous revolution in CPS research and continues to demonstrate great potential for providing solutions to current and impending challenges [18]. This revolution is prominent in vehicular CPS like autonomous vehicles, because of the need to continually make dynamic decisions like lane changing and respond to traffic signs autonomously through image and pattern recognition.

3 Different DL Models in CPSs

We will review the different DL models used in CPSs and discuss how they can be applied to cyber security in this section. Using the DL model, the overall concept of CPS, and the application of DL for CPSs is depicted in Fig. 2.

3.1 Convolutional Neural Networks (CNNs)

CNNs are among the most popular classes of deep neural networks nowadays. The term 'convolution' comes from a linear mathematical operation between two matrices

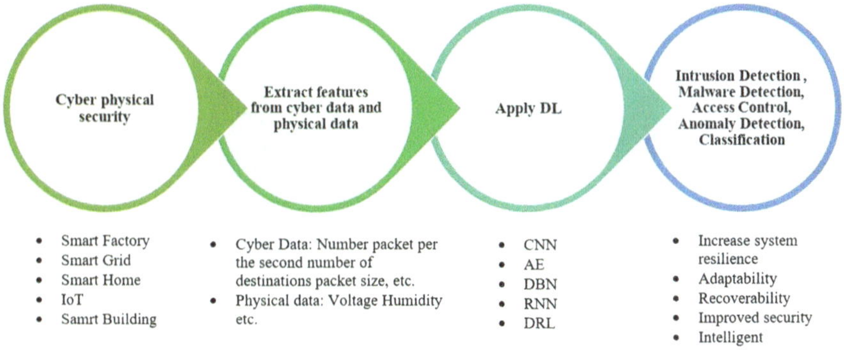

Fig. 2 Application of DL for CPSs [5]

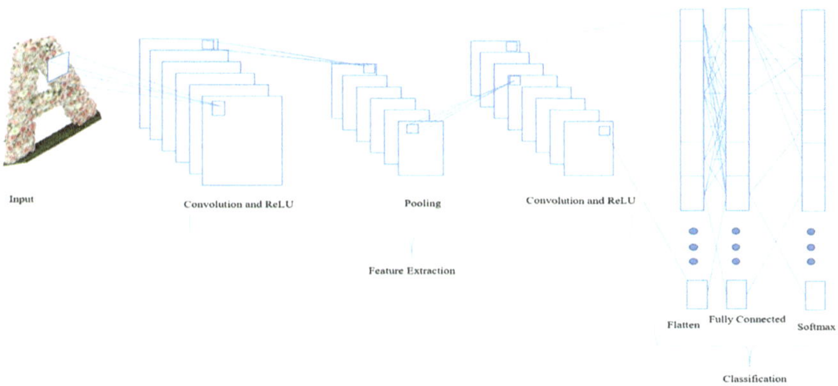

Fig. 3 Typical CNN architecture [21]

called the 'convolution of matrices.' As shown in Fig. 3, a convolutional network has multiple layers, including convolution, pooling, fully connected, and non-linear (activation) layers. Usually, it performs well when used for classification tasks even with raw input. According to [19], the feature map is calculated from the input. Then, promising features are transferred to fully connected layers to perform classification. CNNs were initially developed to process and analyze images, but they were also applied successfully to other types of data, e.g., detecting intrusions. CNNs are often included in IDSs to help extract features from raw data [20].

3.2 Auto Encoder (AE)

An AE is a type of DNNs used to learn proper coding of unlabeled data. The encoding is validated and revised by regenerating the input from the encoding. The AE attempts

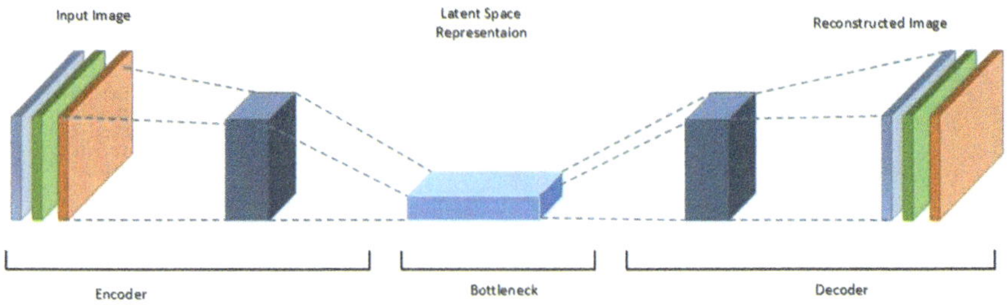

Fig. 4 Structure of AE [23]

to learn a representation (encoding) for a set of data. Typically, this is useful for dimensionality reduction: training the network to capture important data and ignore insignificant ones. As depicted in Fig. 4, an AE consists of two main parts: an encoder that maps the input to the code, and a decoder that maps the code to a reconstructed input. Consequently, the encoder receives the inputs and feeds them to the hidden layers. Through the training procedure of the AE, a code is generated by the encoder and passed to the generator. The generator then reduces the reconstruction errors to refine the learned coding. Thus, task discovery and analysis are carried out using AEs based on their capability to discover what tasks are needed [22].

3.3 Deep Belief Network (DBN)

DBN is formed by a directed acyclic graph with stochastic variables [24]. DBN operates under the principle of greedy selection. There are two critical aspects to consider when designing DBNs: unobserved variables and learning problems. The most significant advantage of DBNs over other DL models is their accurate prediction on unlabeled data. In recent years, DBNs have successfully been applied to a wide range of applications such as image classification, speech recognition, and information retrieval. They have also been successfully applied to natural language processing and cyber-security [25] (Fig. 5).

3.4 Recurrent Neural Network (RNN)

RNNs can be viewed as an enhanced version of feed-forward neural networks, where data is processed in only one direction: from the input layer toward the output layer. Figure 6 depicts the structure of an RNN. An RNN is associated with one or more feedback connections, which function as loop activation, connecting the next layers (nodes) to the previous ones. Such a network is used to perform sequence learning and temporal procedures. The concept of annealing is used to collect and analyze the

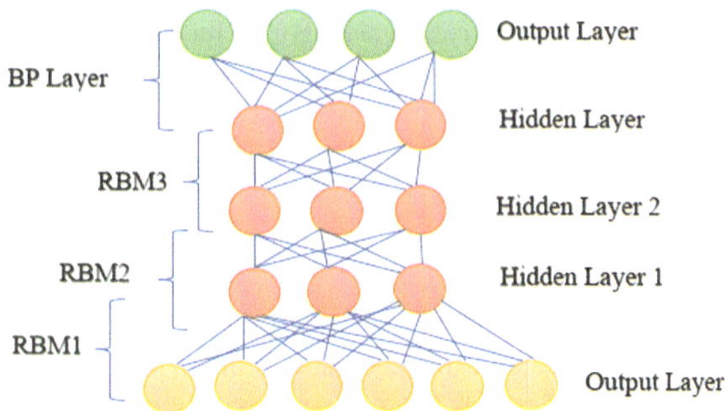

Fig. 5 Structure of the deep belief network (DBN) [25]

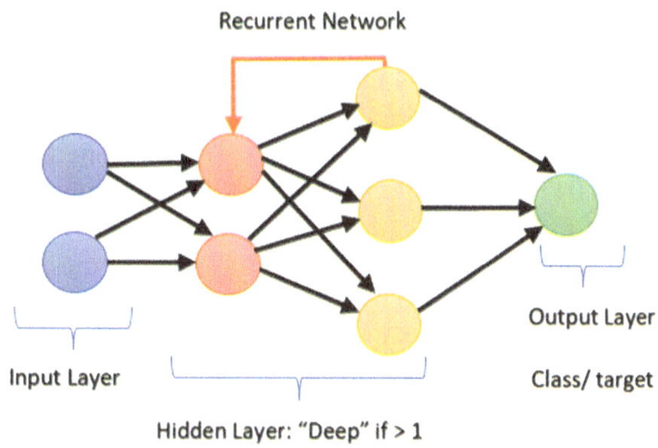

Fig. 6 Structure of recurrent neural network (RNN) [28]

recurrent features. It has been through various studies that these architectures can provide impressive results in various applications, including CPSs security [26, 27].

4 Leveraging DL to Detect Attacks in CPSs

The purpose of this section is to review the state-of-the-art research in the area of cyber-attack detection in CPSs using DL.

4.1 Using Convolutional Neural Networks (CNNs)

A CNN-based model was proposed for cyber-physical security protection in [29] using Siamese neural networks and a few-shot learning model. The proposed model is called the few-shot learning model with Siamese convolutional neural network (FSL-SCNN). The Siamese CNN employs a feature representation optimized for performing task-specific learning to improve the efficiency of learning high-dimensional feature sets. By including three precise losses into the underlying cost function, an improved algorithm is developed for intelligent anomaly detection in industrial CPSs. Based on experiments, FSL-SCNN can provide a significant reduction in false alarm rates (FAR) and F1 metrics for cyber security protection in industrial CPSs. These improvements are made based on two datasets. The first one is a fully labeled public dataset, UNSW-NB15, that was created by the Australian security laboratory for CPS. This dataset is composed of network traffic packets created using IXIA PerfectStrom tool, including realistic modern normal activity and synthetic contemporary attack behavior packets. Moreover, another dataset of few labeled samples used in the experiment is generated in an intelligent CPS for smart manufacturing in which the network transmission packet is collected via the SCADA system and contains a small number of randomly generated abnormally high or low transmission rates signals. The FSL-SCNN is capable of not only separating accurately anomalous signals from standard signals, but also in the few-shot model, it can reduce the false detection rate by utilizing a relative-feature representation scheme and a robust cost function. The F1 score and the FAR are reported as 0.936 and 0.047 respectively. For CPS to detect anomalies effectively, FAR is a critical metric to evaluate.

Another model based on CNN for detecting message injection attacks in vehicular networks is presented in [30]. Since there is currently no security protection for the controller area network (CAN), malicious packets can quickly be injected into the CAN bus via network packets resulting in taking over the vehicle. It is therefore essential to follow the CPSs requirements and be able to detect any malicious package sent to a physical vehicle in the same manner. Several Raspberry Pi devices have been connected to an On-Board Diagnostics (OBD)-II port on an operational passenger vehicle to test the model's effectiveness. The first device acted as a listener, while the second was an attacker. There were four primary types of attack that the attacker exploited: DoS attacks, fuzzy attacks, drive gear spoofing attacks, and engine RPM gauge spoofing attacks. Real-time detection of all attacks in an efficient and effective way poses a considerable challenge. The final result of the model was generated through two blocks of convolutional layers, a pooling, and then a fully connected layer, followed by Softmax activation to make up the final result. There is a sub-dataset for each attack scenario. In one study, CNN outperformed other classifiers, such as Artificial neural networks (ANNs), support vector machine (SVM), long short-term memory (LSTM), k-nearest neighbors' algorithm (k-NN), Naive Bayes (NB), and Decision Trees. As a result, the recording datasets produced for each session comprised 300 injection attacks. Table 2 shows the using four attack scenarios and total messages.

It has been observed that the CNN model made better predictions in terms of the false-negative rate (FNR) and the error rate (ER) where FNR is the fraction of undetected frames that are truly attack frames, and the ER is the fraction of incorrectly classified frames. Based on the CNN model, the results showed that the gear spoofing, the RPM spoofing, the DoS, and the fuzzy attacks had an FNR of 0.06, 0.07, 0.10, and 0.24, respectively. In terms of ER, the CNN model achieved 0.03 for DoS attacks, 0.04 for RPM spoofing, 0.05 for gear spoofing, and 0.18 for fuzzy

Table 2 Attack types and total messages

Total of messages	Type of attack
3,078,250	DoS normal
587,521	DoS attack
3,347,013	Fuzzy normal
491,847	Fuzzy attack
2,766,522	Gear normal
597,252	Gear spoofing
2,290,185	RPM normal
654,897	RPM spoofing

attacks. CNN models were shown to be a promising technique for detecting false message injection attacks in vehicular networks based on experiments performed using empirical data.

A federated DL scheme is also proposed to detect cyber threats targeting industrial control systems [31]. Firstly, the authors presented a model for detecting intrusions using CNN and Gated Recurrent Unit (GRU). Then, the researchers considered a federated learning scenario. Federated learning is an ML technique that trains a model across multiple distributed (decentralized) devices or servers containing local datasets, without exchanging or merging them. They developed a framework for allowing multiple industrial CPSs to build intrusion detection models through federated learning. To ensure the confidentiality of the model parameters, they adopted a communication protocol with the Paillier cryptosystem, a secure communication method. The CPS cyber threat detection framework has been able to detect cyber threats by using federation-based interceptions of data sources and model parameters. Therefore, they proposed DeepFed, based on CNN-GRU, to detect these threats. They have presented the design of a CNN-based parser consisting of a CNN module, a GRU module, an MLP module, and a layer of Softmax. Using a real-world dataset generated by a pipeline system, the researchers evaluated the performance of DeepFed, with 80% of the dataset being used for training and the rest for testing. DeepFed was evaluated with different numbers of local agents, noted by (K). With $K = 3$, DeepFed may reach accuracy, precision, recall and F-score as high as 99.20%, 98.86%, 97.34%, and 98.08% respectively. All measures could achieve the overall metric of 97%. DeepFed was shown to be able to detect multiple types of cyber threats to industrial computer systems due to the experiments that were conducted.

4.2 *Using Auto Encoder (AE)*

In the context of smart power networks, the AE model has been proposed in [32] with the goal of preserving information privacy. The issue of privacy in smart power networks is becoming more prevalent each passing day. It can be difficult to defend

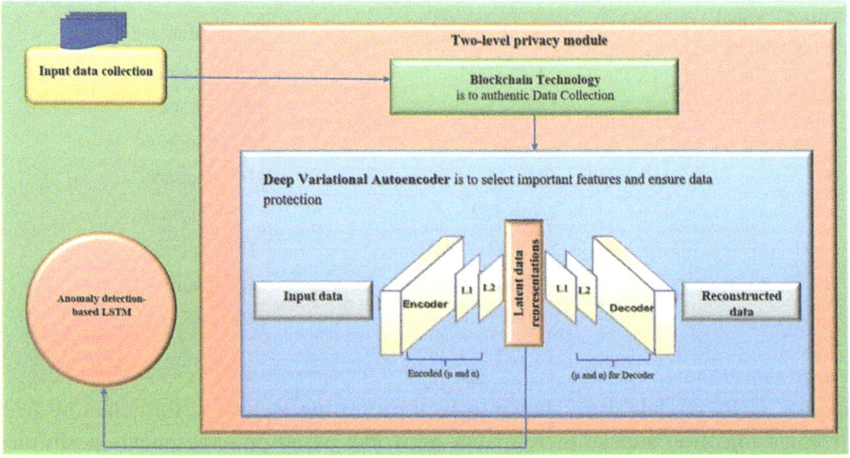

Fig. 7 DL-based blockchain framework for protecting smart power networks proposed in [32]

against inference attacks because smart power networks exhibit characteristics of CPSs, such as monitoring physical processes, closed control loops, attack sophistication, and legacy technologies. On the other hand, Variational AE (VAE) converts the raw data into an encoded format, making it more secure against inference attacks in the classification process. A VAE uses a set of weighted parameters to encode data with a feed-forward network. The proposed model consisted of five input layers, four hidden layers, and one output layer (see Fig. 7). The VAE was evaluated by using two datasets, i.e., the UNSWNB15[1] and the power system datasets [33]. There were 37 scenarios in the power system dataset, including 8 natural events, 28 intrusive events, and 1 no event. The UNSW-NB15 dataset contains a mix of regular and attack records. The performance of the proposed framework was assessed by taking 300,000 randomly selected legitimate and attack observations from each dataset.

Though the VAE was only utilized as a part of the IDS, its abilities were demonstrated when it transformed complex data into a simple format. Concerning the power system dataset, VAE achieved an accuracy of 92.1% and a loss of 0.005, while VAE achieved a precision of 99.8% and a loss of 0.0001 on UNSW-NB15.

It is suggested in [34] that an AE-based approach could be used to detect various cyberattacks in industrial control networks. The web can be accessed through control networks, so many kinds of cyberattacks can occur. The proposed AE consists of an input layer, four hidden layers, and an output layer. The proposed feature space consists of 41 units on the input layer, and the output layer has five divisions for the different types of network traffic. The last hidden layer, which is called Softmax, provided stability to the model. The AE was trained from NSL-KDD[2] dataset. According to early studies, the proposed AE could not detect small classes of attacks

---

[1] https://www.unsw.adfa.edu.au/unsw-canberra-cyber/cybersecurity/ADFA-NB15-Datasets/.
[2] https://www.unb.ca/cic/datasets/nsl.html.

like probe attacks and remote attacks. The stacked AE achieved 97.8% for accuracy over the five categories. The model also achieved an F1 score of 96.8%.

AEs were used to detect cyber-attacks in industrial control systems in [35]. There are various characteristics of this problem, including physical process monitoring, attack sophistication, and legacy technology. Therefore, it was a classification problem within the domain of ML. An algorithm was proposed that uses a 1D CNN. An AE consists of five layers: an input layer, a corruption layer that applies Gaussian noise to the input, a fully connected layer with activation functions, an encoder layer, and a decoding layer. The Secure Water Treatment (SWaT) dataset[3] was used to train the model. The AE trained significantly faster than the 1D CNN in less than half a second. AE's precision, recall, and F1 scores were 89.0%, 82.7%, and 84.4%, respectively. In the end, the AE model proved to be an efficient and effective way to extract useful features.

The idea of detecting cyber-attacks through LSTM autoencoding for autonomous vehicles (AVs) was proposed in [36]. As a result of the communication technologies that are used in antivirus software, they are prone to network attacks such as spoofing attacks and denial of service attacks. These attacks can be detected based on the network traffic. For the purpose of detecting these attacks, LSTM autoencoders were developed. A number of statistics were computed from the network traffic in order to represent the AV activities. Two types of layers were utilized in the neural network architecture: LSTM and fully connected layers. A number of LSTM layers were applied to encode the transformed likelihood stream. An output reconstruction layer was then applied to produce the transformed likelihood stream. As part of the evaluation of the proposed scheme, two datasets were used, namely, the Car Hacking dataset and the UNSWNB15 dataset. Using the Car Hacking dataset, the LSTM based autoencoder, achieved a precision of 99%, a recall of 100%, and the F1 score was 99%. It has been found that according to the proposed scheme, the UNSW-NB15 dataset achieved a precision of 100%, a recall of 97%, and an F1 score of 98%. This indicates that the proposed scheme is able to detect several different types of attack vectors.

4.3 Using Deep Belief Network (DBN)

A DBN-based intrusion detection model was proposed in [37]. Using blockchain-based data transmission and a classification model for CPSs in the healthcare sector, authors proposed an intrusion detection model for secure data transmission. The data collection and processing of data in this model involves various types of sensors, and the intrusion detection model is used to detect intrusions using different types of sensors. Data, which can be processed and transmitted, is integrated with physical processes in CPSs. Considering the fact that medical data in the hands of patients are governed by legal and ethical considerations, cybersecurity is a basic and challenging

[3] https://itrust.sutd.edu.sg/testbeds/secure-water-treatment-swat/.

issue for the healthcare industry. For this reason, it is imperative that the development of the CPSs model for healthcare applications is given specific attention to ensure that the privacy of the users and the security of the data are protected. In some models, multiple share creation (MSC) was used to generate multiple copies of a captured image. Furthermore, blockchain technology has been applied to safeguard the data transmission process between the cloud server and the database. Moreover, a Residual Network (ResNet) based classification model is used to determine the presence of the disease. A validation experiment was carried out on the NSL-KDD 2015 dataset, the CIDDS-001 dataset, and the ISIC dataset to validate the presented model. Using the NSL-KDD2015 and CIDDS-001 models, the simulation results indicated that the presented DBN model was able to detect 98.95% and 98.94% of threats. A more comprehensive analysis of the presented ResNet model shows that it exhibits an appropriate level of classification performance, with a sensitivity of 96.15%, specificity of 98.09%, and accuracy of 98.45%.

A model based on DBN to detect false data injection attacks on the energy internet was also proposed [38]. On the energy internet, they found that it can be quite challenging to detect stealthy cyber-attacks due to the abundance of control signals and meter reading information. This has resulted in the monitoring of physical processes, the establishment of closed loops, and the use of legacy technology to represent the past. The detection problem was formulated as a dual bi-level programming problem with upper and lower bounds since the variations of electric loads can be predicted. To solve the prediction problem, a regression model was used, which had been trained to forecast electric load, based on historical data. The DBN was made of three stacked Random Box Models (RBM) forming six layers: an input layer, four hidden layers, and a logistic regression layer as the output layer.

Data was collected during training to build simulated IEEE 14 and 118-bus systems and the DBN was trained on them. The overall error rate was used as the model's performance metric, and DBN achieved a 2.73% error rate, which was almost 3% lower than SVM as the benchmark model. A recent study shows that when the DBN is used only as a forecasting component of the IDS, it is possible to trade off the DL model with other programming solutions.

It has been suggested in [39] that a model based on DBN can be used to detect false data injection attacks in electric power networks. The CPS characteristics of false data injection detection are mainly based on physical process monitoring, closed control loops, and legacy technology. As the research problem was a classification problem, they used DL to solve it. In order to investigate the detection accuracy, features were automatically generated using DL. There is, however, a proposal to use a DBN model as a baseline that can be compared to an RNN model as well as a graph neural network (GNN). The DBN was created without optimizing any TensorFlow settings, which means that default settings were chosen from the TensorFlow package. Two simulated IEEE 30-bus and the other IEEE 27-bus are used to test the performance of IEEE 118-bus. Using the IEEE-30 bus, the DBN model has been found to achieve 99.39% precision and 98.23% recall, which is very similar to the GNN model and slightly higher than the RNN model. There is one thing to keep in mind, however: the DBN model was consistently better than the RNN model from the IEEE 118-bus, as well as the GNN model. The DBN model has been shown to be reliable, consistent, and has a high rate of stability, as well as the GNN model showing a high rate of potential.

4.4 Using Recurrent Neural Network (RNN)

The use of RNNs has been proposed as a model for detecting cyberattacks on smart grids in [40]. It can be said that smart grids, in addition to being the target of several cyber-attacks, are also the target of data intrusion attacks, denial of service attacks, and so on. DoS attacks, intrusion attacks, data theft, and other sorts of cyberattacks are common against smart grids. A large smart grid network has quite a few characteristics that make it difficult to detect all attacks, including CPS characteristics and legacy technology. Therefore, a vanilla RNN model was trained using the Backpropagation Through Time (BPTT) algorithm. The RNN was evaluated using three datasets, namely, the CIC 2017, Bot-IoT, and power system datasets. The CICIDS2017 dataset includes brute force attacks, botnet attacks, denial of service attacks, website attacks, Heartbleed attacks, and infiltration attacks. There were various security breaches, DoS attacks, and information theft among the Bot-IoT data set. There were injections of data, remote command injections, as well as replay attacks in the power system data set. The CICIDS2017 dataset is made up of 2,830,743 records, the Bot-IoT dataset contains 73,360,900 records, and the power system dataset 78,404 records. A series of experimental outcomes were obtained by analyzing the three respective datasets individually, demonstrating that the RNN model was able to outperform all benchmark classifiers. Using the results from the study, the authors have determined that the false positive rate for RNN classification was 0.00986 for the dataset of CICIDS2017, 0.01281 for the dataset from the Bot-IoT, and 0.03986 for the dataset from the power system. Among the datasets, the RNN model achieved an accuracy of 98.94% for CICIDS2017 dataset, 99.91% for Bot-IoT dataset, and 96.88% for the power system dataset, respectively. Although the vanilla RNN model performed well across all datasets, it remains to be seen how it can be integrated into the DeepCoin of the blockchain component for the detection of fraudulent transactions despite its good performance across all datasets. Nevertheless, this could serve as a starting point for further experiments utilizing blockchain technology as well as DL models.

5 Leveraging RL and DL in Detecting Cyberattacks in CPS

This section reviews how DL and RL can be leveraged in detecting cyber-attacks in the context of CPSs.

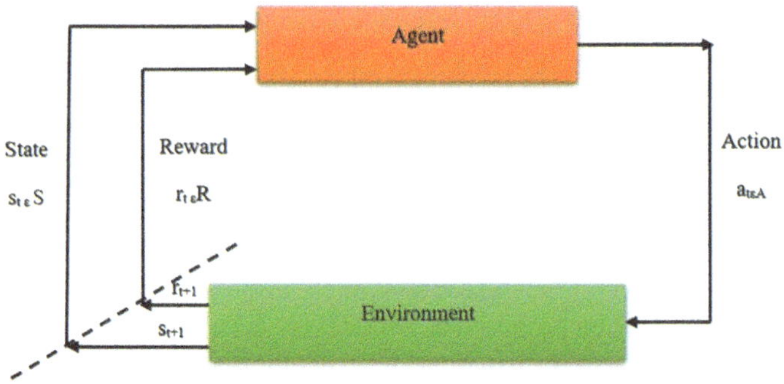

Fig. 8 Interactions between the agent and environment in a RL system [41]

5.1 *Using Reinforcement Learning (RL)*

A RL agent learns how to make decisions through interaction with its environment. During a trial-and-error learning procedure, the agent is rewarded or punished for its performance, respectively. The RL learning algorithms aim is to maximize the cumulative reward overall. In Fig. 8, the two major components of an RL system—the agent and environment—are shown interacting.

The environment represents the external conditions or objects the agent is interacting with. An RL problem also includes a reward signal that represents the environment's feedback upon the agent's actions. Since the agent is attempting to maximize reward through interaction with the environment, it must take advantage of past experiences. On the other hand, the agent needs to explore novel actions (not explored previously) to maximize reward if it wants to choose better actions in the future. Reward is dependent on the agent's current actions and the state of the environment in which it is operating.

To maximize its reward, the agent typically manipulates its policy. An agent's behavior can be predicted by a model of the environment based on the information it has about what happened in each state and how the agent responded to it. The RL represents scenarios in which an active decision-making agent interacts with its environment, where the agent seeks to effectively accomplish a goal without knowledge of the environment [41].

5.2 *Deep Reinforcement Learning (DRL)*

RL and DL are components of DRL. A DL solution incorporates DL into the decision-making process, enabling agents to use unstructured data to make decisions without manually engineering the state space. With DRL algorithms, it is possible to optimize objectives by taking into account significant inputs. Table 3 represents a summary of features of DRL type and their notable methods.

Table 3 Summary of features DRL type and their notable methods [42]

DRL types	Value-based	Policy-gradient	Actor-critic
Features	Compute the value of action given a state Q (s, a)	No value function is needed	Actor produces policy π (s, a)
	Explicit policy learned	Explicit policy is constructed	Critic evaluates action by V (s)
	Sample efficient	Sample inefficient	Often perform better than value-based or policy-gradient methods
Typical methods	Compute value of action given a state Q (s, a)	REINFORCE Vanilla Policy Gradient	DDPG D4PG
	explicit policy learned	TRPO	A3C
	Sample efficient	PPO	UNREAL
Applications	Suitable for problems with discrete action spaces, e.g., classic control tasks: Acrobat, Cart Pole, and Mountain Car as described and implemented in the popular Open AI Gym toolkit	More suitable for problems with continuous action spaces, e.g., classic control tasks described and implemented in the Open AI Gym toolkit: Mountain Car Continuous and Pendulum or Bipedal Walker and Car Racing problems	

In addition to being used for cybersecurity applications, RL has also been applied to several aspects of personal data protection. Cyber security problems are complex and large-scale, and traditional reasoning methods cannot cope with them. The number of connected IoT devices has increased substantially in the last few years, making cyberattacks more complex and numerous. Using deception attacks, launching distributed DoS attacks, infiltrating computer networks, deploying jamming, spoofing, malware, or interfering with a network in an adversarial environment are all examples of how an attacker can inject false data into a CPS. Cyber security topics like multi-agent approaches, combining network-based and host-based intrusion detection, modeling free or modeling-based approaches, and exploring continuous action spaces in cyber environments deserve attention. Aside from human-on-the-loop architectures, deep fakes, poisoning ML, adversarial ML, bit-by-bit distributed systems, and denial-of-service attacks, there is also a discussion of quantum computers cracking encryption algorithms [42].

Researchers have leveraged ML technologies to increase the level of automation of vehicles and make them perform tasks previously performed by humans. The use of sensors and other monitoring devices in vehicles and other infrastructure has become common practice in the past few years. With the help of DRL, the sensor and device data are analyzed for information that is used to make critical decisions on the road. Because of the inherent uncertainty in autonomous driving of vehicles, DRL is used for making decisions such as crossing intersections, changing lanes, controlling speed, and evaluating safety and security. Furthermore, the deep Q-Network and

Q-learning are the most used RL algorithms for research in autonomous vehicles. An interesting research area is to study how DRL could be applied to autonomous vehicles and other forms of CPS. Moreover, DRL is expected to guide other types of decisions as well. However, even though the researchers' findings are theoretical, if they are going to be used in real-life scenarios, it is necessary to do more to guarantee the security and safety of the systems [41]. Access control methods cannot completely thwart adversarial attacks, which is why algorithms must be developed that can operate despite these attacks. Nevertheless, some researchers have argued that the reason there are few research studies that use this model is because there are various challenges to overcome, such as lack of a stationary training set and a distinct right action for every state [43].

In the next few years, we expect to see more research into adversarial attacks and defenses in RL. These developments, however, are dependent on some research findings, as is evident from the research trend presented in this section. First, intermittent attacks that are conducted over a subset of time steps give an adversary even more ability to operate stealthily and efficiently. Secondly, DRL models will continue to be threatened by the ability to entice agents into taking actions leading to adversarial rewards. ATNs and adversarial black-box attacks will extend the frontier of research in DRL. Improving the efficiency of creating adversarial examples such as the ATN will further extend research in DRL. DRL applications such as those for drones, self-driving cars, and other safety–critical systems must address these factors. A comparative analysis of RL and DNN adversarial defense shows that it is still in its infancy [44]. DRL's use in CPS and other systems will however continue to grow, which will draw attention to the issue of defense.

There appears to be more efficiency and practicality in adversarial training for RL than adversarial training for AI that relies on the generation of examples. New defense methods that do not require the generation of examples during training will be more efficient and useful in uncertain environments. It is also important to note that designing systems without a concern for security and adversarial attacks right from the onset has previously presented a challenge in addressing security issues afterwards. In light of this, it is imperative to factor security concerns into the design of systems in order to make them more resistant to adversarial attacks [41].

6 Data Acquisition in CPSs

The acquisition of data is crucial in the process of training DL models. A researcher's effectiveness in finding a solution to a ML research problem is dependent on the quality and quantity of data that are collected. Additionally, data can also influence the predictive model's performance. This is because data is utilized to set up ground truth in supervised learning. Alternatively, researchers could make use of a set of existing datasets collected by other researchers. Table 4 provides a listing of some of existing datasets for cybersecurity research for CPSs.

Table 4 The existing dataset of CPSs

Dataset	Description
SWaT[4]	Data were collected from 11 days of network traffic from a scaled-down water treatment plant. In the first week, no attacks occurred. 36 types of cyber-attacks are listed in this dataset which are among the most prevalent in today's CPS systems
SCADA IDS[5]	Consists of injection of random response packets, hiding the real state of the controlled process, injection of malicious state commands, injection of malicious parameter commands, injection of malicious function code commands, DoS attack, and recon attack
CICIDS2017[6]	This dataset, which identifies the behavior of 25 users based on several protocols, including email protocols, HTTP, HTTPS, SSH, and FTP, was created by making use of several protocols. Among the data, one can also find many records concerning security attacks like Brute Force SSH attacks, Brute Force FTP attacks, Web attacks, Heartbleed attacks, and DDoS attacks
The UNSW-NB15[7]	A software tool referred to as IXIA Perfect Storm is incorporated into the process to create the abnormal and the normal network traffic traces of the dataset. The objective of this tool is to assess the effectiveness and efficiency of NIDSs. It is a tool capable of simulating nine different types of cyber-attacks and its data is updated periodically via a site that provides information on security vulnerabilities
The KDD99 Cup[8]	This dataset is a long version of the DARPA dataset that contains seven weeks of network traffic traces, containing four gigabytes worth of TCP dump data and five million records. It is estimated that 4,900,000 single connection vectors will be used for training KDDCup'99. These vectors contain 41 features that can be classified either as attacks or normal. This dataset contains the following types of security attacks: DoS attacks, User to Root attacks (U2R attacks), Remote to Local attacks (R2L attacks), and Probing attacks. In a U2R attack, an attacker will gain access to the root account of the target host by first gaining access to the user account on the target host. When an attacker performs an R2L attack, he sends data packets to a remote host with the intention of gaining access to it. The other type of attack is a probing attack, which means that the attacker gathers information about a computer network in order to carry out a subsequent security attack on the network
The power system [33]	There are three cyber-attacks that are covered in the blue book: data injection attack, remote command injection attack, and replay attack
The Bot-IoT [45]	A bundle of network traffic logs from an IoT setup, there are three sorts of cyberattacks: infiltration, distributed denial of service (DDoS), and data theft

[4] http://itrust.sutd.edu.sg/dataset/SWaT.

[5] https://sites.google.com/a/uah.edu/tommy-morris-uah/ics-data-sets.

[6] https://sites.google.com/a/uah.edu/tommy-morris-uah/ics-data-sets.

[7] https://www.unsw.adfa.edu.au/unsw-canberra-cyber/cybersecurity/ADFA-NB15-Datasets/.

[8] http://kdd.ics.uci.edu/databases/kddcup99/kddcup99.html.

7 Challenges to Attack Detection in CPSs

In the event of a cyber-attack, the underlying systems could be exposed to potential risks. CPS security is generally addressed by conventional approaches which traditionally address physical and cyber systems separately and are not able to address vulnerabilities that are associated with networks and embedded controllers which are meant to keep track of and control physical processes. Therefore, if the CPS is to be protected against cyber-attacks, it must take a comprehensive approach to security. The literature provides substantial evidence on the importance of protecting systems like these, but there is also substantial evidence that if security is ignored, havoc may ensue [46].

The researchers in [47] used static application analysis in CPS to detect malicious code injection attacks and exploit the possibility of a worst-case execution time. In [46], the authors proposed a lightweight algorithm for matching attack signatures and packet payloads, as well as other techniques that require fewer matching to detect possible attacks. A system for detecting and mitigating sinkhole attacks using IPv6 over LoWPANs is proposed in [48], by combining a trust and reputation system with watchdogs' nodes, to detect and mitigate sinkhole attacks for IoT. A specialized IDS is for Medical CPS using a behavior-rule specification-based method [49]. Behavior rules are transformed into state machines that can detect deviations from specifications of the medical device behavior.

The model proposed in [50] utilizes a mechanism inspired by the Artificial Immune System (AIS) that models' detectors as immune cells classifying any datagram based on matching signatures. New conditions and environments can be monitored, and this can be used in assessing how well the environment is. To learn different variants of the CPS algorithm, code mutations are used in [50]. In [51], it is suggested to equip CPS with automated mechanisms using ML. In this way, they can have self-adaptive mechanisms to deal with anomalies. In [52], it is emphasized that CPS is a typical system with big data, so it is imperative to assess security risks using large volumes of data. As blockchain technology was used in several CPS integrations, including IoT described in [53]. The lack of an integrated security framework poses potential risks to CPS developments [54]. In addition to security, CPS requires intelligence, integration, and cooperation [55].

There are limitations and challenges, such as ineffective IDS, even though there are multiple ideas to detect attacks. The IDS market has a wide range of products, so one can choose from anomaly-based tools [56], behavior-based tools [57], and signature-based tools. However, these products are generally designed to protect IoT-based systems and are not specifically designed to protect CPS systems. Also, CPS systems contain plenty of components and subsystems that will also be susceptible to failures if they are not properly maintained. There is a possibility that the performance of certain sensors deteriorates over time, resulting in incorrect readings. When trying to achieve resiliency of the system, such systems will classify a hardware fault as an attack and take appropriate measures to protect the system from being taken out of action due to this fault. It is also important to use security systems that are trained to identify the correct behavior, as well as hostile patterns in data poisoning attacks.

8 Robust Attacks Detection

Understanding and detecting cyber-attacks is the first step towards developing robust CPS. A reconstruction of attacks is necessary to ensure continuous service of critical infrastructure, as well as to detect attacks quickly. Further, real-life data must be explicitly considered during detection and reconstruction, to account for possible perturbations and/or modeling errors. Recently, it has been discussed to prompt detection and reconstruction using adaptive sliding mode observers paired with parameter estimators and robust differentiators [58]. In spite of the lack of external perturbations, bounded observation errors can be achieved using residual signals. According to [59], it is essential for CPS robustness to be derived from existing results of physical systems. Cyber components of a CPS are particularly crucial when it comes to robust design. Having robust cyber components requires a clear understanding of their behavior. In addition, [60] provides cost functions and transducers. Furthermore, robustness models are discussed from a verification and synthesis perspective. A fast optimization solver able to perform MPC at megahertz rates is described in [61]. It is possible to solve linear-quadratic MPC problems involving inputs and states using various custom computational architectures. Nesterov's fast gradient method is applicable to the architectures, which are suitable for input-constrained problems. A state-constrained problem can be solved using ADMM-based architectures. CPS control must be able to operate reliably despite communication limitations and limited resources. The optimal design of control for arbitrary nonlinear processes is investigated in [62]. An inequality of stability was derived by relating the plant state open-loop growth, the packet erasure probability, and the parameters of the availability model to the plant state open-loop growth. Event-triggered schemes instead of continuous state updates are used by the sensors to reduce communication overhead.

9 Conclusion

With a CPS, physical components, such as manufacturing units in industries, are managed by an automated system. The maintenance system needs to be secured from any type of attack in order to handle manufacturing as well as many control applications. Consequently, CPS security measures are essential. DL plays a key role in detecting security attacks on the CPS system. The purpose of this review was to report such mechanisms and to discuss how well-known DL models such as CNN, RNN, and AE have been applied for various tasks in CPS. Moreover, DRL (combination of DL and RL) has demonstrated a great ability to provide solutions to current and future challenges.

The fields of cybersecurity and AI are all set to continue evolving. Research is focused on the challenges encountered in applying these technologies to real-life problems, as well as the opportunities presented by them. This research area is becoming more and more interdisciplinary, so there are some emerging technologies or innovations that are contributing to the actualization of AI-driven cybersecurity, particularly in CPS. Following are some of the problems that we face today in the area of CPS:

- This groundbreaking research in DRL has opened promising directions of research in a wide range of current and envisioned applications, including navigation, robotics, air traffic control, and defense. In the near future, DRL will be of increasing interest in the cybersecurity space.
- As attacks have become more sophisticated and large-scale, the defenses themselves must be improved as well. To tackle this problem, we can explore the possibility of multiagent DRL.
- The majority of DRL algorithms used for cyber defense are model-free that require a large quantity of training data. Due to real-life cyber security problems, real-life training data is extremely difficult to obtain. Researchers often use simulators to validate their proposed approaches. Despite this, these simulators fail to capture the real complexity and dynamics of the underlying cyberspace forms the IoT. Model-based methods are more appropriate and practical than model-free methods when training data is limited since they can be scalable. Exploring and integrating model-based and model-free DRL methods for cyber defense would be an interesting future study.

References

1. Orbis Research (2020) Global cyber physical system market 2020 by Company, regions, type and application, Forecastto 2025 Orbis Research. Retrieved from https://www.orbisresearch.com/reports/index/global-cyber-physical-system-market-2020-by-company-regions-type-and-application-forecast-to-2025
2. Check Point Software (2021) Cyber attack trends mid-year report
3. Yaacoub J-PA et al (2020) Cyber-physical systems security: limitations, issues and future trends. Microprocess Microsyst 77:103201
4. Luo Y et al (2021) Deep learning-based anomaly detection in cyber-physical systems: progress and opportunities. ACM Comput Surv (CSUR) 54(5):1–36
5. Wickramasinghe CS, Marino DL, Amarasinghe K, Manic M (2018) Generalization of deep learning for cyber-physical system security: a survey. In: IECON 2018—44th Annual conference of the IEEE industrial electronics society, pp 745–751. https://doi.org/10.1109/IECON.2018.8591773
6. Mitchell R, Chen I-R (2014) A survey of intrusion detection techniques for cyber-physical systems. ACM Comput Surv 46(4)
7. Xin Y et al (2018) Machine learning and deep learning methods for cybersecurity. IEEE Access 6:35365–35381. https://doi.org/10.1109/ACCESS.2018.2836950
8. Goodfellow I, Bengio Y, Courville A (2016) Deep learning. MIT Press

Security and Privacy of IoT Devices for Aging in Place

Noel Khaemba, Issa Traoré, and Mohammad Mamun

Abstract The rising cost of elderly living and care facilities prompts for other solutions for the elderly people where aging in place is one of the ways to solve this issue using emerging technologies centered around smart IoT devices. To ensure security and privacy for a smart home for aging in place, different aspects of the IoT devices have to be considered. This chapter seeks to provide a categorical review and analysis of age-tech IoT device technologies, and discuss the underlying security and privacy challenges and available solutions.

Keywords Security · Privacy · Internet of things · Ageing in place · Dataset · Agetech IoT Solution · Sensor · Raspberry Pi · Cloud storage · Actuator system · Threat · Vulnerabilities · Mitigation strategy · Machine learning

1 Introduction

The global life expectancy index is expected to continue increasing as years go by [1]. For instance, in the UK, it is predicted that by 2035, the number of people aged 65 and above will be approximately 16.9 million [2]. When people live longer there will be significant increase in health issues for the elderly. Also, with the increase in the elderly population, the societal cost for care facilities and services for them will go up [3]. Many families cannot afford care givers or home nurses and it is challenging for family members to take care of the elderly as they might have to stop working. This creates the need for Ageing in Place (AIP) whereby the elderly population

N. Khaemba (✉) · I. Traoré
Department of Electrical and Computer Engineering, University of Victoria, Victoria, BC, Canada
e-mail: noelk@uvic.ca

I. Traoré
e-mail: itraore@ece.uvic.ca

M. Mamun
National Research Council Canada, Ottawa, NB, Canada
e-mail: Mohammad.Mamun@nrc-cnrc.gc.ca

I. Traore et al. (eds.), *Artificial Intelligence for Cyber-Physical Systems Hardening*, Engineering Cyber-Physical Systems and Critical Infrastructures 2,
https://doi.org/10.1007/978-3-031-16237-4_8

prefer staying in their homes instead of care facilities for elderly people despite their health or mobility challenges [1]. To enable AIP, technological solutions based on smart Internet of Things (IoT) devices come in handy as they help meet the needs of the elderly people [4]. Data collected from these devices can for example show the person's routine and tell when there are inconsistencies and prompt for emergency services [2]. IoT refers to the connection of objects or material devices with the Internet at the center, and the devices have sensors [1]. In the context of elderly people, the IoT smart technology is referred to as Age-Tech which is technology that supports or provides solutions for them and their caregivers.

Despite the growing number of Age-Tech solutions currently available, limited attention has been paid to the security and privacy risks posed by these technologies. While most of these solutions face the same security and privacy threats as general purpose IoT platforms, those threats are exacerbated by the life criticality of most devices and the fact that many elderly not being technology savvy opens up the door for increased vulnerability.

The objective of this chapter is to present a categorical review and analysis of Agetech IoT solutions and a discussion of corresponding security and privacy challenges and approaches. Emphasis is placed on outlining solutions based on artificial intelligence (AI) and machine learning (ML), and corresponding datasets and classification models.

The rest of the chapter is structured as follows. Section 2 presents the major categories of devices used for AIP and outline corresponding use cases. Section 3 discusses the common threats and vulnerabilities faced by aging in place technologies, by highlighting threats specific to the elderly and threats common to the broad IoT user population. Section 4 discuses the use of artificial intelligence for agetech security, by outlining relevant datasets and approaches. Section 5 makes concluding remarks.

2 Review of IOT Devices for Aging in Place

In this section, we present different IoT device types applicable for ageing in place and discuss relevant usage scenarios.

2.1 Device Types

Activities of daily living are important when monitoring an elderly person's wellness. These are day-to-day life or self-care activities like bathing, working, eating and cleaning. There are three types of sensors for daily living activities as follows [5]:

- **Physical Environment Based Sensors** like pressure, proximity, Radio Frequency Identification (RFID), WIFI and Zigbee sensors. They identify action through items or their connections with objects.
- **Wearable Sensors or Body-appended Sensors** like accelerometers, pedometers, goniometers, gyroscopes, electromechanical switches and inertial sensors. Thse devices are meant to provide physiological and biomechanical information. These help distinguish human body exercises in day-to-day living.
- **Other Activity Recognition Sensors** include camera footage or audiovisual arrangements. These devices help recognize mortal movements or social actions.

There are three main uses for the aforementioned sensor technologies in ageing in place, that is detection of emergencies, monitoring health status and notifying medical experts of changes in ones health [1]. Other uses include automating home maintenance and daily tasks, support for transport and navigation and enabling connection or communication with different social networks and caregiving communities.

Dependent on the usage scenario, different combinations or grouping of sensors could be used. For instance, Pandya et al. collected activity data in elderly peoples' homes by deploying an experimental setup involving 5 sensor types and an ESP8266 WiFi micro-controller (MCU) placed together with each sensor in different locations of the house [5]. The five sensors used are as follows:

- **Passive Infrared (PIR) Sensor** that detects human motion.
- **Infrared (IR) Sensor** for motion detection in the surroundings in case there is an obstruction. The infrared radiation detects how far an object is and when there is movement.
- **Hall Magnetic Sensor** that detects the presence and magnitude of a magnetic field.
- **Pressure Mat Sensor** that detects pressure on surfaces like on a bed or sofa.
- **Temperature–Humidity Sensor (dht-11)** that detects the temperature and humidity in a given space.

These sensors are considered to be of low cost in the market, ambient and non-intrusive to the residents. The ESP8266 WiFi micro-controller helps pre-process and transfer data to a Raspberry Pi B+ server. These logged data and system logs are periodically sent to a cloud storage for backup. It is mentioned that this system had shortcomings, for instance, there was inaccurate data generated by wrongly calibrated sensors.

2.2 Use Cases

There are different sensing and actuator systems in homes for elderly people. These sensing technologies can help monitor wellness, social, seasonal and weather related characteritics of routine tasks. The data generated from such monitoring can be used to construct baseline models by learning daily activity patterns of the elderly and identifying unusual or anomalous patterns. An alerting and anomaly detection system that can automatically monitor the activities of the elderly and identify unusual occurrences is very helpful as it can be used to remotely monitor their health and wellness. Such detection system can record contextual data of the various areas of an elderly person's smart home and if it detects any anomalies, it alerts the caretakers [5].

For instance, Fattah et al. [3] implemented a prototype AIP system that issues to old adults reminders for medication in different ways including wristwatch, smartphone, and home appliances such as lights and speakers. It is expected that the proposed infrastructure, which consists of the aging-in-place service platform and service composition tool will be able to contribute to reducing the increasing healthcare cost for elderly people and also give informal caregivers including family members and friends peace of mind.

3 AIP Threats and Vulnerabilities

Succcessful attacks against AgeTech can have far-reaching impact, such as the following:

1. **Misdiagnosis**: This is, for example, when an attacker accesses the IoT device data and modifies it, that means the information is incorrect and can lead to wrong diagnosis of an elderly state of health.
2. **Exposure of Personal Identifiable Information (PII)**: This is a risk to themselves and their family for example when financial information is exposed especially because smart devices are interconnected. When a loophole in one device is exploited, sensitive information can be gathered from many devices.
3. **Privacy Intrusion**: The personal privacy of the aged person is interfered with when their sensitive information is exposed to unauthorized people. This may include records of health information and daily living patterns.

Threats against AIP can be categorized into threats specific to such environment and relevant threats that are common to any environment.

3.1 Threats Speficic to AgeTech Environment

Frik et al. highlighted in [6] the fact that most IoT devices are not designed specifically for the elderly population and because of their novelty, this poses security and privacy concerns. To understand the security and privacy threats faced by agetech, the authors cconducted semi-structured interviews with 46 elderly people. This helped determine their security and privacy concerns, misconceptions that they may have and it helped identify some common threat models and remediation processes. The authors noted that finance, health and the living situation affect the older adults interaction or decisions around smart devices [6]. When the IoT devices security risks are mitigated, it is noted that this causes technical difficulties for the aged, making it difficult for them to use the devices and some may decide to avoid smart devices entirely. Hence, there is a need to educate the elderly on how to use the devices and at the same time make sure there is improved usability of the device without compromising on its security [6].

Some studies, such as the work by Fournier et al. [7] have a different perspective, as they argue that difficulties in the use of technology is not age-related but it is a result of peoples' education and work experience [7]. Currently, most elderly people did not use as much technology during their working life so this impacts their ability to use IoT devices. Smart devices used at home are more susceptible to security breaches like unauthorized access compared to those in a hospital or nursing home that are in a controlled environment [7].

Elderly people illnesses like mild cognitive impairment (MCI) do pose major cybersecurity risks in aging in place [8]. MCI is characterized by decline in cognitive function, for example, challenges in coming up with words, frequently forgetting to take medicine or go to appointments and impaired geographic orientation. Some of these illnesses normally come with age and affect ones life. In severe cases, one may suffer from dementia. The use of smart devices and internet connection is beneficial to the elderly, however, when they suffer from such illnesses, they are more prone to security risks, for example, they can easily be defrauded.

Mentis et al. [9] conducted a study to gain more insight on the decision making process for online security and safety for elderly people with MCI and their caregivers. The authors conducted semi-structured audio interviews, recorded and transcribed in verbatim. The semi-structured interview consisted of a list of questions and topics of discussion that the interviewer doesn't have to strictly follow. Instead they allow room for the conversation to go to different trajectories that are meaningful. The interviews were conducted in the homes of elderly couples' and involved explaining the project to the participants and signing a consent document. The data was analyzed using thematic analysis whereby the data was coded and divided into themes. Mentis et al. concluded that although the aged people expressed interest in taking part in making security and privacy decisions, there is a lack of options to safeguard and enforce shared-decision making with their caregivers. Majority of caregivers make the decisions by themselves without involving the elderly people [9].

Beside MCI, there are other aged population health factors that can cause security risks with age-tech. For instance,

1. **Impaired Vision**: As people age, they are likely to have decline in their visual clarity. This makes them vulnerable to security attacks, for example, an attacker can replace their smart device with a malicious device without them noticing because they're not very alert or can't see so well.
2. **Parkinson's Disease**: The Parkinson's disease is characterized by decline in coordination, trembling or shaking of the jaw, arms, legs, head, stiffness and challenges

in walking. This is a disease that comes as one ages, it may hinder their capability to manage the security and privacy of their smart devices because they may focus more on their well being and what easily works for them.

3.2 Common Threats

There are several threats common to all IOT devices. Obviously, these threats are applicable for AgeTech, and in some cases, they are aggravated when applied for the elderly.

IoT devices are sometimes left unattended to; this physical exposure makes it easy for an attacker to take the device, extract information, modify its programming or even replace it with a compromised device that the attacker controls [10].

According to Kaspersky [11], most users of smart devices use them without changing the default credentials. Therefore, one of the attack strategies from hackers is guessing passwords since most smart devices use default passwords. In order to collect data and analyse attacker activities on smart devices, a honeypot was setup, and it was discovered that attackers use infected IoT devices to launch malicious activities or perform Distributed Denial of Service (DDoS) attacks [11]. These are some of the attack strategies that infiltrate devices and applications used in aging in place.

Older adults have decline in their abilities or capabilities as they continue aging, they may also not have much technical know-how of especially novel technologies and may be reluctant to learn as well. There are several other attacks they are likely to face, most of which are old types of attack techniques, including the following.

3.2.1 Phishing and Spamming

The aged population is an exposed target to social engineering [12]. They are more likely to be tricked using social engineering tactics and also considering that attackers come up with new strategies every other day, the elderly people are less likely to be informed or remain updated on newer cybersecurity risks. They can easily be tricked to give their personal information, for instance, through a phone call or text message.

3.2.2 Scareware

This in most cases is a pop-up, email or message that asks the user to take immediate action or else they face certain consequences. This may scare off an elderly person who can click on it and give their PII to an attacker without their knowledge.

3.2.3 Access Tailgating

This is piggybacking where a person who is not authorized to enter a place follows the elderly person behind and enters their home without their knowledge. Since elderly people are generally slow in their movement it is possible for this to happen.

3.2.4 Baiting

This is where something desirable is presented to the user, for example, asking them to listen to a video or participate in a competition to win money. Considering that some elderly people have financial challenges, they may be tempted to try this and end up accessing malicious sites from an attacker.

3.2.5 Smart Device Theft

With living in place, most homes are not well monitored, for example, with CCTV cameras everywhere so the smart devices with sensitive information can be stolen.

3.3 Device Specific Threats

In this section, the focus is to understand how specific devices work and the threats they may face. IoT devices for aging in place can be divided into two categories, that is domestic smart devices that help with tasks in the home and health equipment that help monitor or support different health factors. Some of the IoT devices used by the elderly population include self-learning stove alarm, lights, IoT controlling temperature, smart barcode that can be connected to a smart oven to identify cooking temperatures, smart doorbell, smart speakers, and reminders for medication using wristwatch or smartphone [3]. Smart devices can also be divided into three categories as below.

3.3.1 Wearable Devices Like Smartwatches

The attacks that wearable devices face include man in the middle (MITM) attack whereby the data payload is intercepted and modified. The solution to this is performing encryption [13]. There is also malicious code injection where the attacker takes full control of the sensor node. Other attacks include firmware attack, mole attack, mule attack and user account hijacking attack.

3.3.2 Motion Sensor Devices

These include devices like gyroscope, accelerometer and linear acceleration sensors. One attack is denial of service whereby an attacker injects the sensor with a signal to make it unavailable to the user. Other attacks include eavesdropping, transmission of malicious commands, keystroke inference and false sensor data injection [14]. There is a need for a context-aware model that can learn the device's and user's behavioural patterns and adapt to it.

3.3.3 Environmental Sensors

These include light sensor, camera, temperature sensor, audio sensor, and proximity sensor. The attacks they are likely to face are similar to those of the motion sensor devices [14].

Table 1 lists different types of sensors and possible attacks.

Table 1 : Sensors and possible attacks

Type of attack	Wearable sensors	Motion sensors	Environmental sensors
Firmware attack	✓	✗	✗
Mole attack	✓	✗	✗
Eavesdropping	✗	✓	✓
Transmission of malicious commands	✗	✓	✓
Keystroke inference	✗	✓	✓
False sensor data injection	✗	✓	✓
Mule attack	✓	✗	✗
User account injection attack	✓	✗	✗
MITM attack	✓	✗	✗
Denial of Service (DoS)	✗	✓	✓
Malicious code injection	✓	✗	✗

3.4 Mitigation Strategies

Previous research has focused on providing recommendations that will enhance adoption of IoT devices by the elderly in their home while helping mitigate some of the security threats.

For instance, to counter the impact of MCI on AgeTech security, Mentis et al. [9] recommended having shared decision making systems. Research shows that more often than not, the caregivers make decisions without consultation yet the elderly people would like to be involved in the decision making [9]. This method is not feasible because there isn't proper safeguarding of a range of options for the caregiver and the elderly person to choose from. There is also need for a balance in reinforcing security while at the same time allowing the elderly person to enjoy social interactions or communication through smart devices. There should be a legal way for caregivers and the elderly person to establish some form of trust such that the privacy and security of the data generated from the IoT devices is safeguarded.

For better joint decision making, a person-centered approach is recommended where the care-recipient has a voice in decision making with discussion with their family or caregivers. They are able to preserve their safety and autonomy and not feel like decisions are being made for them. It is noted that, even among healthy people, making choices in cybersecurity is still challenging. There is insufficient consumer involvement in security and privacy decisions whereby most of the time they're presented with long privacy policy documents that most people don't read [9]. In order to recognize a security threat, a healthy level of cognitive ability is necessary. It is already hard for elderly people without cognitive decline to notice phishing scams and so it is even worse for those suffering from cognitive impairment like MCI [9].

To help strenghthen security and privacy, Barralon et al. recommended using new cryptographic schemes such as Identity Based Encryption [4]. They also highlighted two crucial goals that age-tech must help to achieve, that is optimizing home delivery of elderly care so that they do not have to be admitted to hospitals or such occurrence can be postponed and when hospitalization occurs they can be discharged faster because they have a home with smart devices to support them [4].

Almeida et al. proposed in [8] a new way of collecting and managing data for elderly peoples' activities. This involves gathering data across several cities from heterogeneous devices, which allows for comparison of behaviour patterns of the aged in the different locations. The proposed system is able to serve multiple cities concurrently and it consists of Linked Open Data paradigms and IoT. These are in two layers: personal data capturing system and personal data storage and management system. In this solution, the data is integrated across cities and each city can have a different level of data abstraction bringing about an endpoint that can be queried and deductions made from the data.

Research on AgeTech security is at an early stage. Most of the proposals on this topic focus on general guidelines as illustrated above. An area full of promise for AgeTech security is using machine learning and artificial techniques in mitigating relevant threats and vulnerabilities. We discuss in the next section efforts in this direction.

4 Using Machine Learning and AI Models

Data is one of the most valuable ingredients of AI and applied machine learning techniques. In this section, we present existing datasets and proposals towards addressing AgeTech security challenges using ML/AI techniques.

4.1 Available Datasets

4.1.1 Anomalous Activity in IoT Networks Dataset

Ullah et al. [15] reviewed different intrusion detection datasets and identified several weaknesses.

For instance, they noted that DARPA98/99 datasets which are widely used as network intrusion datasets have often been criticized for containing many redundant records. The KDD99 dataset which is a widely used dataset for intrusion detection has TCP attributes, 5 million data instances and more than 20 different types of attacks but failed to provide information about IP addresses. They observed also that University of Twente's flow-based intrusion detection labelled dataset by Sperotto that was the first publicly available dataset in this category has a high ratio of malicious instances since the network traffic was collected from a honeypot. This affects the learning algorithm to be biased in the direction of the most frequent class and reduces the ability of the detection algorithm from learning the least common class.

Having determined dataset weaknesses, they came up with a new dataset called IoTID20 for anomalous activity detection in IoT networks [15].

4.1.2 Multi Sensor Dataset of Human Activities

This dataset which consists of human activities data based on multiple sensors deployed in a smart home was collected by Chimamiwa et al. [16]. The authors recommended that this dataset can be used for habit or activity pattern recognition using machine learning methods. The dataset consists of records of information from sensors of activities of a person living in a smart home. 6 types of sensors were used, including temperature-humidity, motion, light, contact, pressure and smart plug sensors. 23 sensors were deployed in the smart home, including 3 force sensing resistors,

6 passive infrared, 3 reed switches, 1 temperature and humidity sensor, 3 mini photocell light sensors, and 7 smart plugs [16]. A WiFi micro-controller programmed to read data and send it to a central database was attached to each of the sensors. The message queuing telemetry transport (MQTT) communication protocol was used in gathering and sending data for storage on a central database server. The data was collected over a period of six months at a frequency of 1 Hz. The user information collected from the smart home environment include indoor movements, pressure applied on the couch or bed, use of the TV, fridge and stove, or the use of electrical appliances like microwave, coffee maker, dishwasher, sandwich maker or washing machine. Three tables were used to record the data as explained below.

- **Sensor Table**—information such as the id, datatype of the sensor reading, and sensor name.
- **Sensor-Sample-Int Table**—integer measurements from motion, pressure, light, and contact sensors. Contact and motion are binary, either 0 or 1.
- **Sensor-Sample-Float Table**—floating point datatype from temperature and humidity sensor and the smart plugs.

The intention behind choosing the type of sensors used in this experiment is so that they can capture the resident's indoor activities of daily living (ADL) like sleeping, sitting, cleaning, cooking and bathing. Two major categories of sensors were used in this experiment.

I. Ambient Sensors

These are sensors that capture information concerning the atmosphere or air. The following ambient sensors were used:

- Mini Photocell Light Sensor—used to measure light intensity, for example, when the stove or TV is switched on.
- PIR Motion Sensor (Grove)—used to detect presence of the resident in a specific area of the home.
- Si7021 Temperature and Humidity Sensor BoB—used to measure temperature and humidity of the air particularly in the bathroom.

II. Object Sensors

These are sensors that are attached to objects in the house so that they can capture data concerning those specific items. The following object sensors were used:

- Force Sensing Resistor—attached to the couch, bed and weight scale so as to tell the quantity of pressure applied as the resident interacts with the item.
- Reed Switch—attached to a room's door or the door of the fridge to detect when the door is opened or closed.
- TP-Link Wi-Fi Smart Plugs HS110—They have energy monitoring ability, therefore they can measure the amount of current being used by the different electric appliances like the electric kettle, washing machine, microwave, coffee maker or vacuum cleaner

4.1.3 Open Smart Home Dataset

The open smart home dataset was collected by Schneider et al. by deploying a smart home with different sensors that can measure temperature, humidity and brightness. The authors didn't provide information about the specific type of sensors used. They placed the sensors in the kitchen, bathroom, toilet, and three different rooms (room1, room2, and room3) in a flat. The dataset is a time series dataset which consists of records of readings from the sensors and the respective time the measurement were made [17]. The dataset has the following types of measurements:

- ThermostatTemperature—this is temperature of the air measured using 2 thermostats installed in room 2. The thermostat is mounted to the radiator. The data type is float and degree Celsius is the unit.
- Brightness—a luminance sensor is placed in each room to measure brightness. The data type is float and lux is the unit.
- Humidity—a humidity sensor is mounted on the wall in each room to measure the percentage of relative humidity in the atmosphere. The data type is integer.
- Temperature—a temperature sensor is placed in each room to measure the indoor air temperature. The data type is float and degree Celsius is the unit.
- SetpointHistory—Set point is the desired temperature value that is set on the remote or which a thermostat has been initially set to. The data type is float and degree Celsius is the unit.
- OutdoorTemperature—This is the temperature outside the building that is obtained from a virtual weather service. The data type is float and degree Celsius is the unit.

4.1.4 Smart Building Dataset

This is a time series dataset collected by Hong et al. [18]. The dataset was collected from 4 floors of a building in UC Berkeley consisting of 51 rooms and involving 255 sensors in different areas. 5 measurement types were collected from each room over a period of one week. These measurements include CO2 concentration, PIR motion sensor data, room air humidity, luminosity and room temperature.

The data consists of the sensor reading together with the time the measurement was collected in Unix Epoch Time. In every 5 seconds, all the sensors are sampled except for the PIR motion sensor that is sampled every 10 seconds. The PIR sensor helps determine whether there's room occupancy or not because it measures infrared (IR) light radiating from items in its field of view. When the PIR value is zero, it shows that the room is empty and when it is non-zero, it indicates that the room is occupied [18].

4.1.5 Smart Home Dataset With Weather Information

This dataset was collected by Zainab et al. [19] to study IoT devices used at home together with the weather information. The goal of the experiment was to determine the trustworthiness of readings or information collected by smart home devices. The dataset was collected over a period of one year at a frequency of one minute. The data includes the following areas and appliances: fridge, dish washer, living room, barn, microwave, garage door, well, wine cellar, solar and home office. It consists of the following weather features: temperature, pressure, cloud cover, wind speed, icon, precipitation probability, humidity, dewPoint, precipitation intensity, visibility, windBearing, and apparent temperature.

4.1.6 REFIT Smart Home Dataset

This dataset was collected by Firth et al. [20] from 20 smart homes in the UK over a period of slightly more than 3 years. For each home, a building survey was conducted to get information about the occupancy, construction materials, energy services and building geometry . Below are the sensors that were set up in each home:

- Current Cost Mains Clamps—this is for measuring the electrical power load used in the home.
- Replacement Gas Meters—for measuring the household's mains gas consumption.
- Hobo U12 or Hobo Pendant Sensors—for measuring light level, temperature in a room and relative humidity.
- iButton Temperature Sensors—for measuring radiator surface temperature.
- CurrentCost Individual Appliance Monitors—for measuring plug electrical power loads.
- RWE Smart Home Devices—these include smoke sensors, exterior and interior motion detectors, programmable thermostatic radiator valves and door and window opening sensors.
- British Gas Hive Programmable Thermostats—it helps one to control the heat in a home remotely for example using a smartphone application such that you don't have to heat the home when it's empty and can schedule to have it warm when you come back.

4.1.7 Dataset from Assembled IoT Testbed

This dataset was collected by Anthi et al. [21] from an IoT testbed that consisted of different smart home devices. These include Belkin NetCam camera, Amazon Echo Dot, TP-Link NC200 Camera, Samsung Smart Things hub, TP-Link Smart Plug and British Gas Hive that is connected to two sensors: a window/door sensor and motion sensor, and Lifx Lamp. Various attacks were executed and in order to record the network traffic and generate and save logs automatically, a laptop was connected to the network.

A tcpdump tool was used to collect the data that consist of benign data points and malicious data points. Two weeks worth of malicious data and three weeks of benign data was collected from the IoT testbed so that the data can conform to other comparable research datasets [21]. The cyber-attacks that were executed include Man-In-The-Middle (MITM), Spoofing, Denial of Service (DoS), Replay and Reconnaissance. The methods used to implement the attacks in each of the attack categories are as below.

- Reconnaissance—NMAP scan, iot-scanner
- Denial of Service (DoS)/DDoS—hello flood attacks, UDP/TCP flood
- MITM—Burpsuit, SSL Strip, Ettercap
- Spoofing—ARP, DNS
- Replay—mitmframework suite.

4.1.8 Observations

A few key observations can be made about the above datasets. These datasets mainly focus on sensors that can provide information which can determine patterns in terms of the well-being of the resident in the smart home. However, considering aging in place, data about health is very crucial, so measurements for, for example, blood pressure, and heart rate are necessary.

In the Multi Sensor Dataset of Human Activities [16], only one home was considered. Having data collected from several homes will provide stronger ground to investigate and address a much broader range of research challenges. The Open Smart Home Dataset [17], mainly has sensors for temperature, humidity and brightness which implies that there isn't much variety in the type of sensors. For aging in place, it is good to consider more sensor types to measure a broader variety of features.

The Smart Building Dataset [18] has a luminosity sensor which is quite unique compared to the sensors mentioned in other resources. The luminosity sensor is considered to be a sophisticated light sensor that measures both visible light and infrared to better approximate the response of the human eye.

The Dataset from assembled IoT testbed [21], focused on collecting network data of IoT devices in order to come up with supervised Intrusion Detection System for Smart Home IoT Devices. This is different from the other datasets that focused on collecting sensor recordings.

4.2 *Existing ML-Based Proposals*

4.2.1 Anomaly Detection Using ML for AIP

The two traditional ways of remote health monitoring are use of body sensors to measure different body parameters and use of systems with sensors in a smart home that send reports to health specialists. The first option requires the elderly person to be involved which can be tedious and inconvenient. The second approach relies on medical experts to review the reports which is time consuming and costly. Therefore, there is a need to develop an automatic anomaly detection and alert system that requires minimal human interaction. To achieve this, Pandya et al. proposed a behavioural pattern recognition approach that is based on reinforcement learning and machine learning models [5]. This health monitoring system is able to collect and automatically process data, then generate reports and alerts that are sent to caretakers who can reach out to medical experts when need be. Pandya et al. analysed monthly data from two smart homes for elderly people where they collected over 5000 test samples and used it to validate the accuracy of the proposed system [5].

Piau et al. [22] performed a study to assess the relevance of the automatic sensor alerts based on the feedback from the elderly people, their caregivers and professionals who supported the elderly people. This was a way to help evaluate the performance of the real-time behavioral anomaly detection algorithm. They also further analysed the sensor capability to tell when the elderly person has a major medical issue and decline in cognitive abilities. In their research the participants were 25 elderly people above 75 years of age that lived by themselves in their homes. They did a follow-up for 6 months. The study targeted frail elderly people who are at the verge of losing or already lost their cognitive abilities but they are able to walk without assistance.

Piau et al. [22] stated that as much as technology has been effectively used to manage many medical issues, there are fewer solutions that focus on monitoring intrinsic capabilities of the elderly for AIP. There is a lack of comprehensive solutions that allow monitoring the various health aspects when aging in place, instead, many of the used smart devices overlap in function and there is a tendency to use several tools when it would be better to have a consolidated one. The research targeted different non-intrusive sensors and one particular sensor that stood out is the push-button on a bracelet or pendant that could be used in case of an emergency to trigger an alert. They proposed an alert system that is triggered by the AI algorithm when it detects significant changes in indoor activities of daily living (ADL) patterns. In the first month, the algorithm learns from the activity data recorded by the sensors, thereafter it detects behavioral anomalies in real-time. They noted that the main setback to high adoptability of alert systems is because of factors like false positive alarms and challenges in integrating them into the already complex health care systems.

What stands out from Piau et al.'s research is that the proposed solution caters to the needs of the elderly people living by themselves, it is unobtrusive, it incorporates assessments from end users, it uses low-cost sensors that require minimal maintenance and the AI algorithm continually improves its performance based on the feedback from end users. However, they didn't provide detailed information about the AI algorithm. Therefore it is difficult to assess independently the effectiveness of the proposed solution.

Shahid et al. [23] performed a study whose objective was to develop a model that can learn the older people's day-to-day routine, detect when there are anomalies and generate real-time alerts to their relatives. The study involved setting up in elderly peoples homes off-the-shelf sensors and IoT devices including smart water meters, motion sensors and wall plugs. The study consisted of recording meals and bathroom actvity from 9 apartments.

The authors developed a data analytics architecture for detecting deviations in behavioural patterns using a statistical-based anomaly detection model model. They performed the trial experiment in 9 actual elderly peoples' apartments for 64 days. Thereafter, more data for analysis and modeling was collected for 2 years. In the proposed statistical model, which is based on the Chebyshev's inequality, anomalies were identified when the duration of an activity in each room was outside the limits of two standard deviation from the mean $(\mu - 2\sigma,\ \mu - 2\sigma)$, where μ and σ stand for the mean and standard deviation, respectively. 75% confidence interval was used for the tested data. The mean (μ) and standard deviation(σ) were estimated over hourly time periods for 80 to 355 days depending on the apartment. They also developed an SMS-based notification and user evaluation service that the users could use to rate the service and overall they reported to have had a positive user experience. The user and caregivers could also get notifications for both normal and abnormal daily activities. In future work they recommend using reinforcement learning and more contextual sensors.

Aran et al. [24] performed a study whereby their objective was to observe the daily activities of seniors, and to identify the corresponding patterns and detect outliers. They collected a dataset from 40 households of seniors based on pressure, motion and door sensors. They used a probabilistic spatio-temporal model to interpret daily behaviour and anomalies were singled out using a cross-entropy measure. In creating the model, the assumption is that the location of the elderly person at a given time is depended on the hour of the day. The behavioral model captures the conditional probability of being in a certain location at a given hour which helps to provide a summary of daily activity patterns for each user. The four locations considered are living room, bathroom, bedroom and outing which is when a senior leaves the household. Aran et al. also used k-means clustering method from which they observed that there were two clusters of behavioral patterns among the elderly people whereby most of them spend the day in the living room and slept in the bedroom at night while the others did not use their bedroom to sleep at night. Therefore, for the latter group, the bed sensor may not collect much information. The cross-entropy measure is low when the activities are normal which implies that the learned behavioral model is doing a good job in making predictions but when there are anomalies, they noted the cross-entropy is high which means that the model did not make accurate predictions. For future work, they suggested that there is a need to have anomalies ground truth data with labels as this will help validate the anomaly detection model performance.

Artola et al. highlight the fact that there is not much research that focuses on well being and healthcare anomaly detection. Therefore in their study they suggest developing anomaly detection system for detecting deviations in the behavioural patterns of elderly people. This is to help improve the older persons' wellbeing and early recognition of health issues like cognitive decline. Thw authors collected data from different sources including ADL records from sensors or devices and electronic personal records. They used eHealth sensors, wearables, mobile applications, assistive robots and IoT enabled devices as tools to support the wellbeing of the elderly and gather data. The data which is unlabelled and unstructured is analysed by an anomaly detection system (ADS) that learns activity patterns and can tell when there are abnormal cases or outliers. The ADS consists of several AI methods that are statistical based, rule based, nearest neighbour based, clustering based and classification based. To arrive at a decision in a given instance, they apply the algorithm with the optimal results. The ADS classifies the behaviour or activities of each elderly person using the three colours of traffic lights; green meaning everything is normal, yellow meaning there is something suspicious that needs follow-up while red means there is an anomaly, which must be addressed as soon as possible. These results are presented to the caregivers and health professional via a dashboard. The study is ongoing; in [25], the authors presented a prototype of the project and indicated that in the next phase o f the work, they intend to perform validation of ADS using data collected from a nursing home and a smart living environment for older adults. The shortcoming of this study is that the authors did not provide an elaborate discussion of the AI models used.

Lentzas et al. [26] explored different studies about human activity recognition (HAR) and anomaly detection among elderly people and provided a review of the various frameworks used. They noted that the commonly used unobtrusive sensors include smartwatches or smart bands, smartphones, ambient sensors for force, temperature or humidity and RFID devices. Although there is an increasing focus on using machine learning methods in HAR, many existing approaches are based on statistical models. A few approaches have embraced machine learning main techniques such as decision tree, random forest, hidden markov models, rule based models, support vector machine, recurrent neural networks, fuzzy logic, markov chains, etc. Lentzas et al. recommend further investigating the use deep learning and recurrent neural networks (RNN) as they showed promising results in behavioral analysis and HAR. The most used performance metrics include accuracy, recall score (True Positive Rate (TPR)), precision score (False Positive Rate (FPR)) and F-score. Lentzas et al. noted that as much as recall and precision scores provide a strong way of evaluating anomaly detection model performance in terms of correct and false alarms, an improvement can be made such that the system response time is incorporated in coming up with the overall score. This is especially because time is of essence in case of critical health anomaly.

It can be observed from the above studies and the literature that although there are many papers on anomaly detection for agetech, to our knowledge, none of the proposed approaches address anomalies arising from malicious activities. The existing approaches are focused on anomalies from a well being or health perspective. For example, when sensor values are extremely low or high compared to baseline, it means the elderly person has a health emergency and not that it could be a security breach. There is a gap in the research in identifying and addressing anomalies in aging in place stemming from malicious activities, which is the essence of anomaly detection in cybersecurity.

4.2.2 Security and Privacy for AIP Smart Devices

Ulah et al. [27] used TensorFlow neural network to detect malware files and pirated software targeting IoT devices. Specifically, a novel methodology that involves color image visualization and convolution neural network (CNN) was used to detect malware. In the proposed approach, the files with malware are converted into colour images which give better visualization features, then these are passed through a CNN. It is noted that this combined method produces better results than when using the conventional models. The proposed method focuses on achieving high accuracy, low computational cost and reduced overhead [27].

In processing the data, the malware binary files are converted into gray-scale images, then feature extraction techniques are applied to classify the type of malware. Feature reduction is used to reduce the number of features which in turn improves the classification. The deep learning algorithms performed better than the shallow machine learning algorithms because they can use filters to automatically reduce noise [27].

Gochoo et al. [28] encourage the use of environmental sensors instead of cameras because this preserves the privacy of the elderly person. They recommend using Robots to carry out daily living activities and have sensors integrated with them.

Gochoo et al. focus on reviewing the various technologies and smart home solutions that are currently being used for the betterment of the quality of life of the elderly. They also point out technologies that are privacy-preserving. They noted that from the previous papers they reviewed, the research did not incorporate all aspects that form a complete smart home. They stated that such a smart home should be non-intrusive, have easily accessible emergency features, incorporate fall detection, assist in activities of daily living (ADL), provide companionship for less anxiety or loneliness and incorporate entertainment and gaming. They recommended that to increase unobtrusiveness, the use of wearables smart devices and cameras should be kept to a minimum. The use of robots to assist the elderly person with activities like sitting and standing or bringing the medication was encouraged. Therefore, technology that incorporates robots along with sensors would be highly effective. They also argued that previous studies did not incorporate data security and privacy which is an important factor because it affects the user's trust in the systems [28]. Therefore, it needs to be given very high priority.

5 Conclusion

There is an increasing societal cost for public healthcare services for the elderly worldwide due to their increasing population. Therefore, aging in place is meant to help cater to the needs and support the elderly to age comfortably in their homes without having to pay for elderly living facilitites or being disconnected from their communities and family.

We have provided a review of the different IoT technologies used in AIP, the different types of smart devices used and the type of data collected from sensors and IoT devices. We have also highlighted the security breaches that are likely to happen in an AIP environment, and discussed anomaly detection solutions offered in previous studies and machine learning methods implemented.

There are gaps in aging in place research for resources that address how to handle or manage cybersecurity and ethical issues in age-tech [7]. It is important to preserve the security and privacy of the elderly people using smart devices considering that these devices have their personal data. For them to fully trust the device systems and utilize the different functionalities, the systems have to be secure and unobtrusive to protect their privacy. There are large amounts of data generated in the use of IoT devices for age-tech. In reference to the past few years, the data is increasing at a high magnitude. The use of artificial intelligence and machine learning requires focus on the security aspect of AIP data. There can be serious implications in terms of safety and health of the elderly if data privacy and protection is not taken care of [7]. Therefore, there is a need to put more focus on security and privacy in AIP and have an anomaly detection system that can establish when there is a security breach.

Acknowledgements This project was supported in part by collaborative research funding from the National Research Council of Canada's Aging in Place Program.

References

1. Carnemolla P (2018) Ageing in place and the internet of things—how smart home technologies, the built environment and caregiving intersect. Visual Eng 6:1–16
2. Merry H (2022) IoT: keeping the elderly independent at home. IBM Business Operations Blog. https://www.ibm.com/blogs/internet-of-things/elderly-independent-smart-home/
3. Fattah S, Sung N, Ahn I, Ryu M, Yun J (2017) Building IoT services for aging in place using standard-based IoT platforms and heterogeneous IoT products. Sensors 17:2311
4. Barralon P, Charrat B, Chartier I, Chirie V, Fico G, Guillen S, Homehr N, Kamenova R, Lamotte F, Peine A (2019) IoT for smart living environments: recommendations for healthy ageing solutions. In: Alliance for internet of things innovation
5. Pandya S, Mistry M, Kotecha K, Sur A, Ghanchi A, Patadiya V, Limbachiya K, Shivam A (2021) Smart aging wellness sensor networks: a near real-time daily activity health monitoring, anomaly detection and alert system. In: Proceedings Of second international conference on computing, communications, and cyber-security, pp 3–21

6. Frik A, Nurgalieva L, Bernd J, Lee J, Schaub F, Egelman S (2019) Privacy and security threat models and mitigation strategies of older adults. In: Fifteenth symposium on usable privacy and security (SOUPS 2019), pp 21–40
7. Fournier H, Molyneaux H, Kondratova I, Ali N (2019) Privacy by design and cybersecurity for safe, effective and reliable home health care for aging in place. In: International conference Europe Middle East & North Africa information systems and technologies to support learning, pp 442–450
8. Almeida A, Fiore A, Mainetti L, Mulero R, Patrono L, Rametta P (2017) An IoT-aware architecture for collecting and managing data related to elderly behavior. Wirel Commun Mobile Comput
9. Mentis H, Madjaroff G, Massey A, Trendafilova Z (2020) The illusion of choice in discussing cybersecurity safeguards between older adults with mild cognitive impairment and their caregivers. Proc ACM Hum-Comput Interact 4:1–19
10. Abomhara M, Køien G (2015) Cyber security and the internet of things: vulnerabilities, threats, intruders and attacks. J Cyber Secur Mobility 65–88
11. Kaspersky (2019) IoT under fire: Kaspersky detects more than 100 million attacks on smart devices in H1 2019. www.kaspersky.com. https://www.kaspersky.com/about/press-releases/2019_iot-under-fire-kaspersky-detects-more-than-100-million-attacks-on-smart-devices-in-h1-2019/
12. Frumento E (2019) Cybersecurity and the evolutions of healthcare: challenges and threats behind its evolution. In: M_Health current and future applications, pp 35–69
13. Liu J, Sun W (2016) Smart attacks against intelligent wearables in people-centric internet of things. IEEE Commun Mag 54:44–49
14. Sikder A, Petracca G, Aksu H, Jaeger T, Uluagac A (2021) A survey on sensor-based threats and attacks to smart devices and applications. IEEE Commun Surv Tutorials 23:1125–1159
15. Ullah I, Mahmoud Q (2020) A scheme for generating a dataset for anomalous activity detection in IoT networks. In: Canadian conference on artificial intelligence, pp 508–520
16. Chimamiwa G, Alirezaie M, Pecora F, Loutfi A (2021) Multi-sensor dataset of human activities in a smart home environment. Data Brief 34:106632
17. Schneider G, Rasmussen M, Bonsma P, Oraskari J, Pauwels P (2018) Linked building data for modular building information modelling of a smart home. In: 11th European conference on product and process modelling. CRC Press, pp 407–414
18. Hong D, Gu Q, Whitehouse K (2017) High-dimensional time series clustering via cross-predictability. Artif Intell Stat 642–651
19. Zainab AS, Refaat S, Bouhali O (2020) Ensemble-based spam detection in smart home IoT devices time series data using machine learning techniques. Information 11:344
20. Firth S, Kane T, Dimitriou V, Hassan T, Fouchal F, Coleman M, Webb L (2022) REFIT smart home dataset. Figshare. https://repository.lboro.ac.uk/articles/dataset/REFIT_Smart_Home_dataset/2070091
21. Anthi E, Williams L, Słowińska M, Theodorakopoulos G, Burnap P (2019) A supervised intrusion detection system for smart home IoT devices. IEEE Internet Things J 6:9042–9053
22. Piau A, Lepage B, Bernon C, Gleizes M, Nourhashemi F (2019) Others real-time detection of behavioral anomalies of older people using artificial intelligence (The 3-PEGASE Study): protocol for a real-life prospective trial. JMIR Res Protoc 8:e14245
23. Shahid Z, Saguna S, Åhlund C (2022) Detecting anomalies in daily activity routines of older persons in single resident smart homes: proof-of-concept study. JMIR Aging 5:e28260
24. Aran O, Sanchez-Cortes D, Do M, Gatica-Perez D (2016) Anomaly detection in elderly daily behavior in ambient sensing environments. In: International workshop on human behavior understanding, pp 51–67
25. Artola G, Carrasco E, Rebescher K, Larburu N, Berges I (2021) Behavioral anomaly detection system for the wellbeing assessment and lifestyle support of older people at home. Procedia Comput Sci 192:2047–2057
26. Lentzas A, Vrakas D (2020) Non-intrusive human activity recognition and abnormal behavior detection on elderly people: a review. Artif Intell Rev 53:1975–2021

27. Ullah F, Naeem H, Jabbar S, Khalid S, Latif M, Al-Turjman F, Mostarda L (2019) Cyber security threats detection in internet of things using deep learning approach. IEEE Access 7:124379–124389
28. Gochoo M, Alnajjar F, Tan T, Khalid S (2021) Towards privacy-preserved aging in place: a systematic review. Sensors 21:3082

Detecting Malicious Attacks Using Principal Component Analysis in Medical Cyber-Physical Systems

Wei Lu

Abstract The recent rise of medical cyber-physical systems has rapidly changed the current healthcare industry, its widespread use in hospitals, however, has also paved a way for a large number of cybercriminal activities targeting these networked devices, raising serious security and privacy concerns when healthcare professionals deal with sensitive and life-critical medical information. Existing security solutions in this domain are mainly prevention-based and are highly insufficient due to the power consumption and costly resources when implementing computationally expensive solutions. As an alternative, we propose in this paper an anomaly detection system based on principal component analysis to assure the security of networked medical devices. We evaluate our approach with a publicly available dataset collected in a real-time medical cyber-physical system testbed network and the results show the proposed approach successfully detects malicious attacks with a high detection rate and an acceptable low false alarm rate.

Keywords Malicious attack · Principal component analysis · Medical cyber-physical system · Anomaly detection · Intrusion detection system · Medical features

1 Introduction

The medical cyber-physical system (mCPS) refers to a connected networking infrastructure of medical devices, software applications, healthcare information systems, and digital health services, where the connected medical devices create, collect, analyze and transport health data information or medical images to either a cloud computing facility or internal servers via the healthcare provider networks. The recent rise of mCPS has rapidly changed the current healthcare industry, where the use of medical wireless sensors (e.g. smart pacemaker, smart blood glucose meter),

---

W. Lu (✉)
Department of Computer Science, Keene State College, The University System of New Hampshire, Keene, NH, USA
e-mail: wlu@usnh.edu

I. Traore et al. (eds.), *Artificial Intelligence for Cyber-Physical Systems Hardening*, Engineering Cyber-Physical Systems and Critical Infrastructures 2,
https://doi.org/10.1007/978-3-031-16237-4_9

actuators, and medical surgery robots not only improves patients' quality of health but also enables personalized health services (e.g. Cardio-Pulmonary Resuscitation) [1–5]. In addition, health data collected by mCPS devices are stored and transmitted to advanced cloud computing platforms such as IBM Watson [6], AWS [7], and Microsoft Azure cloud services [8], from which professionals in the medical domain can explore patients' health-related data further to make accurate health prescriptions and decisions [9–11]. In this context, the mCPS has been proven to be a game-changer for the healthcare industry, heading us to a future of smarter and more accurate diagnoses with fewer mistakes and lower costs of care.

On the other hand, the widespread use of mCPS devices in hospitals, however, has also paved the way for a large number of criminal activities to thrive. In March 2019, the FDA issued warnings on dozens of implantable cardioverter defibrillators that could be affected by various malicious attacks such as eavesdropping, message alteration, fake data injection, ransomware, and distributed denial of service, leading to a compromise of patient security, safety, and availability of critical healthcare systems [12–15]. Due to poorly implemented security mechanisms, potential adversaries can easily obtain remote control or hijack a smart medical device by using malware or botnet (e.g. Echobot, Mirai, Emotet, Reaper, Necurs, and Gamut), and then manipulate the sensitive data by injecting fake health data or cause malfunctions by flooding the resource-constrained mCPS network with a large number of illegitimate requests (e.g. manipulating a smart pacemaker to give an abnormal shock to patients) [16–18]. This jeopardizes and threatens patients' lives, thus hindering the wider deployment of mCPS devices [19–21]. Therefore, to achieve the maximum possible benefit and leverage the full potential of mCPS devices in healthcare, there is an urgent need, for novel security mechanisms to preserve the security of mCPS devices.

Existing security mechanisms in this domain include dedicated password-based authentication, anomaly detection, access control mechanisms, encryption, and trust-based security management for wearable, implantable, environmental, and portal medical devices [22–30]. Such prevention-based security solutions are highly insufficient mainly because (1) they can be easily circumvented by using widely available vulnerabilities of embedded operating systems and wired/wireless communication protocols deployed in mCPS devices; (2) mCPS devices are usually considered resource-constrained devices; implementing the computationally expensive security solutions inside the mCPS devices are challenging due to their limitations on power consumption and costly resources. As an alternative solution, we propose in this research a novel intrusion detection system to assure the security of mCPS devices and to detect malicious attacks from inside and outside of the healthcare context [31–34].

The rest of the paper is organized as follows. Section 2 introduces the concept of intrusion detection systems including both misuse detection and anomaly detection techniques. Section 3 presents the proposed detection approach based on the Principal Component Analysis (PCA). Section 4 is the experimental evaluation for our detection model with a publicly available dataset called WUSTL-EHMS-2020 in which both the network traffic data and patients' healthcare data are collected in a real-time Enhanced Healthcare Monitoring System (EHMS) testbed network [35]. Finally, in Sect. 5 we make some concluding remarks and discuss the future work.

2 Concepts of Intrusion Detection Systems

Traditionally, intrusion detection techniques are classified into two categories: misuse (signature-based) detection and anomaly detection. In misuse detection, the attacks are detected by matching actual behavior recorded in audit trails with known suspicious patterns [36, 37]. While misuse detection is fully effective in uncovering known attacks, it is useless when faced with unknown or novel forms of attacks for which the signatures are not yet available [38]. Moreover, for known attacks, defining a signature that encompasses all possible variations of the attack is very challenging [39, 40]. Any mistakes in the definition of these signatures will increase the false alarm rate and decrease the effectiveness of the detection technique [41, 42].

Different from misuse detection, anomaly detection is dedicated to establishing normal activity profiles for the system [43, 44]. It assumes that all intrusive activities are necessarily anomalous [45]. Anomaly detection studies start by forming an opinion on what the normal attributes of the observed objects are and then decide what kinds of activities should be flagged as intrusions and how to make such particular decisions [46–49]. A typical anomaly detection model is illustrated in Fig. 1. It consists of four components, namely data collection, normal system profile, anomaly detection, and response. Normal user activities on mCPS devices are obtained and saved by the data collection component. Specific modeling techniques are used to create normal system profiles. The anomaly detection component decides how far the current activities deviate from the normal system profiles and what percentage of these activities should be flagged as abnormal. Finally, the response component reports the intrusion and possibly corresponding timing information.

The primary advantage of anomaly detection is its capability to find novel attacks; as such it addresses the biggest limitation of misuse detection. Therefore, our proposed approach is based on the anomaly detection framework as illustrated in Fig. 1 in which the PCA technique is used to reconstruct patients' normal behavior profiles.

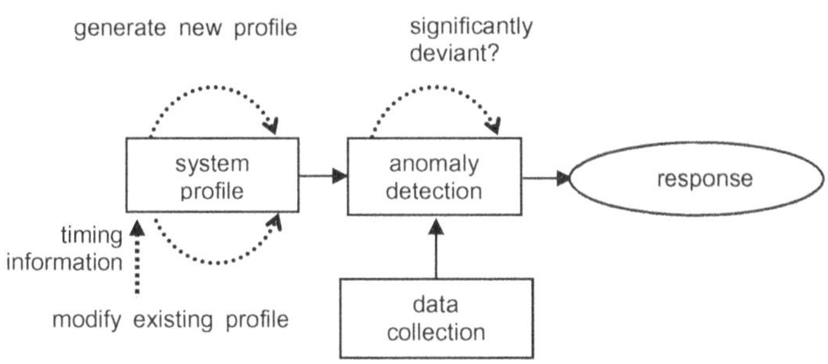

Fig. 1 An mCPS anomaly detection system

3 PCA Based Anomaly Detection

Principle Component Analysis (PCA) is one of the most popular and effective methods of lossy data compression that has been wildly used in the data science domain as a technique to reduce multidimensional data sets to lower dimensions for analysis. PCA uses an orthogonal transformation to convert a set of possibly correlated variables into a set of new variables of linearly uncorrelated, where the greatest variance by any projection of the data comes to lie on the first coordinate (called the first principal component), the second greatest variance on the second coordinate, and so on [50]. These new variables called Principal Components (PCs) are smaller than the original set of variables but also retain most of the sample's information. PCA is mainly used to reduce the dataset to a lower dimensions space and is computationally inexpensive. When applying it for detecting anomalies, we use inverse transformation to reconstruct the original dataset from the PCs. By observing how much the difference of reconstructed data deviated from the original data the anomaly is then identified.

Given A to be an $m \times n$ matrix of m observations and m × n f_i in an n-dimensional space in which $m > n$

$$A = \begin{bmatrix} f_{1,1} & f_{2,1} & \cdots & f_{m,1} \\ f_{1,2} & f_{2,2} & \cdots & f_{m,2} \\ \vdots & \vdots & \ddots & \vdots \\ f_{1,n} & f_{2,n} & \cdots & f_{m,n} \end{bmatrix}$$

where

$$f_i = \left(f_{i,1}, f_{i,2}, \ldots, f_{i,n}\right) \quad 1 \leq i \leq m$$

by using the singular value decomposition (SVD), we have three matrices of U, $\sum$ and W, such that

$$A^T = U \sum W^T$$

where U is an $n \times m$ matrix in which the columns are the left singular vectors of A, $\sum$ is a $m \times m$ rectangular diagonal matrix whose diagonal contains the singular values of A (i.e. eigenvalues λ_i where $\lambda_1 \geq \lambda_2 \geq \ldots \lambda_n \geq 0$) and W is a $m \times m$ matrix in which the columns are the right singular vector (eigenvectors) of A.

The next step is to mean the center of the dataset A from which we have a new dataset N whose mean is zero:

$$N = A - \begin{bmatrix} \mu_1 \ \mu_1 \ \cdots\cdots \ \mu_1 \\ \mu_2 \ \mu_2 \ \cdots\cdots \ \mu_2 \\ \vdots \ \vdots \ \cdots\cdots \ \vdots \\ \mu_m \ \mu_m \ \cdots\cdots \ \mu_m \end{bmatrix}$$

where

$$\mu_k = \frac{1}{n}\sum_{i=1}^{n} f_{i,k} \quad 1 \leq k \leq m$$

based on N and W, the corresponding value of the observation in the new coordinate system of principle components can be derived in the following formula:

$$P^T = W^T \times N$$

where the matrix P is the final data set in the principle components feature space with data items in rows and dimensions along with columns.

4 Experimental Evaluations

The WUSTL-EHMS-2020 dataset is a publicly available dataset developed by the Washington University in St. Louis for the purpose of evaluating the performance of applying machine learning algorithms for detecting intrusions in the medical CPS. The data is collected in an in-door testbed network called Enhanced Healthcare Monitoring System (EHMS), where four components are included, namely medical sensors, gateway, network, and control with visualization. The medical sensors are used to collect patients' biometrics data and then send them to the gateway, where the network traffic data are also captured. Next both network traffic data and patients, biometrics data are then sent to a central server through switch and router. The front-end clients then retrieve data information from the server for visualization and display them to medical professionals. Two types of man-in-the-middle attacks are simulated in this environment, namely spoofing and data injection. The spoofing attack, representing a typical attack violating the patients' data confidentiality, is used to sniff the packets transmitted between the gateway and the central server. The data injection attack, representing a typical attack violating the data integrity, is used to modify the packets in real-time during data transmission. The dataset includes 41 features among which 32 features are collected from network flow metrics, 8 features are collected from patients' bodies via medical sensors and one binary feature is about the label of data record where a value of 1 indicates an attack and 0 means normal. The details on feature names and their descriptions are illustrated in the following Tables 1, 2, 3, 4 and 5.

Table 1 Nominal network flow features

Feature name	Description
SrcAddr	Source IP address
DstAddr	Destination IP address
SrcMac	Source hardware MAC address
DstMac	Destination hardware MAC address

Table 2 Float based network flow features

Feature name	Description
sIntPkt	Interpacket arrival time (millisecond) from source to destination
dIntPkt	Interpacket arrival time (millisecond) from destination to source
sIntPktAct	Active interpacket arrival time (millisecond) from source to destination
dIntPktAct	Active interpacket arrival time (millisecond) from source to destination
SrcJitter	Source jitter (millisecond)
DstJitter	Destination jitter (millisecond)
Dur	Transaction record total duration
Rate	Number of packets per second

The total number of records we have in the dataset is 16,318 of which 2,046 are attacks. When preparing the data, we split the dataset into a training set including 13,054 records of which 1639 are attacks, and a test dataset including 3264 records of which 407 are attacks. Figure 2 illustrates some examples of the descriptive statistics on features of the dataset.

PCA approach essentially reduces the data dimensions while minimizing reconstruction error, giving us the flexibility to reconstruct our original features from the newly created features with an error as little as possible. In this case, it is important to define an anomalous score so we can calculate how anomalous each reconstruction is. The more anomalous the reconstruction of the data, the more likely the data record is malicious with an assumption that the attack is rare and looks very different from the majority of normal data records. As a result of our approach, we define the anomalous score for each data record as the sum of the squared differences between the original data matrix and the reconstructed matrix using the PCA algorithm. This sum of squared differences is then scaled by a max–min range of the sum of the squared differences for each entry in the entire dataset so all the anomalous scores will be within the range of zero to one, where zero means normal records and one means anomaly with the highest probability to be attacked.

To explain intuitively the effectiveness of our approach we use the traditional confusion matrix to evaluate the performance as illustrated in Table 6. True positives (TP) is the number of records in which we predict the record is attacked and it does the attack. True negatives (TN) is the number of records in which we predict the record is normal and it does the normal. False positives (FP) is the number of records

Table 3 Integer based network flow features

Feature name	Description
Sport	Source port number
Dport	Destination port number
SrcBytes	Number of bytes from source to destination
DstBytes	Number of bytes from destination to source
SrcLoad	Source to destination bits per second
DstLoad	Destination to source bits per second
SrcGap	Missing bytes from source to destination
DstGap	Missing bytes from destination to source
sMaxPktSz	Maximum packet size for traffic transmitted from source to destination
dMaxPktSz	Maximum packet size for traffic transmitted from destination to source
sMinPktSz	Minimum packet size for traffic transmitted from source to destination
dMinPktSz	Minimum packet size for traffic transmitted from destination to source
Trans	Aggregation packets count
TotPkts	Total transaction packets count
TotBytes	Total transaction bytes
Load	Total transaction bits per second
Loss	Packets retransmitted or dropped
pLoss	Percent of packets retransmitted or dropped
pSrcLoss	Percent of packets retransmitted or dropped from source to destination
pDstLoss	Percent of packets retransmitted or dropped from destination to source

Table 4 Integer based medical features

Feature name	Description
SpO2	Blood oxygen
Pulse_Rate	Pulse Rate in BPM
SYS	SYStolic blood pressure
DIA	DIAstolic blood pressure
Heart_rate	Heart Rate in Beats Per Minute (BPM)
Resp_Rate	Respiration Rate in BPM

Table 5 Float based medical features

Feature name	Description
Temp	Temperature in degrees Celsius
ST	Electrically neutral area between ventricular depolarization (QRS complex) and repolarization (T wave) in millivolts (mv)

	SrcBytes	DstBytes	SrcLoad	DstLoad	SIntPkt	DIntPkt	SrcJitter	DstJitter	sMaxPktSz	dMaxPktSz
count	16318.000000	16318.000000	1.631800e+04	1.631800e+04	16318.000000	16318.000000	16318.000000	16318.000000	16318.000000	16318.000000
mean	496.650264	187.077706	2.118406e+05	7.102435e+04	10.946755	8.515423	32.025841	8.546796	310.000429	65.999632
std	28.584642	18.688525	7.942988e+04	4.530811e+04	101.230484	52.504560	1266.374694	49.667244	0.020708	0.046970
min	310.000000	120.000000	0.000000e+00	5.074470e+02	0.875000	0.730500	0.000000	0.675500	310.000000	60.000000
25%	496.000000	186.000000	1.990535e+05	6.635500e+04	3.792667	2.226000	2.814472	2.171500	310.000000	66.000000
50%	496.000000	186.000000	2.366790e+05	7.889300e+04	4.191333	2.562250	3.053852	2.505750	310.000000	66.000000
75%	496.000000	186.000000	2.615570e+05	8.719300e+04	4.984167	3.136375	3.680555	3.086000	310.000000	66.000000
max	2298.000000	882.000000	1.134000e+06	3.938000e+06	9497.338000	2445.732000	66099.194140	2268.882000	311.000000	66.000000

Fig. 2 Descriptive statistics on features in the dataset

in which we predict the record is attack and it does the normal. False negatives (FN) is the number of records in which we predict the record is normal and it does the attack. Precision is when our system predicts the record is attack how often the prediction is accurate where we use the following formula to calculate it:

$$Precision = \frac{TP}{TP + FP}$$

and Recall is when our system predicts the record is attack how many this kind of true attack the system can catch over the total number of actual attacks in the dataset, i.e.

$$Recall = \frac{TP}{TP + FN}$$

Figure 3 illustrates the precision-recall curve showing a trade-off between precision and recall in which the best precision our approach achieves is 88.5% where 354 attacks are correctly identified when choosing a cut-off value of 400 predicted attacks.

Next, we conduct two more evaluations with the same PCA based approach, one is based on the patients' biometrics only (i.e. Biometrics) and the other one is based on the combination of network flow metrics and patients' Biometrics data (i.e. Netflow + Biometrics) in which the best precision for Biometrics and the combination is 16.3% and 59.5%, respectively, as illustrated in Table 7. Figures 4 and 5 illustrate the precision-recall curve accordingly.

Table 6 A confusion matrix

Actual value	Predicted value	
	Attack	Normal
Attack	True positive	False negative
Normal	False positive	Ture negative

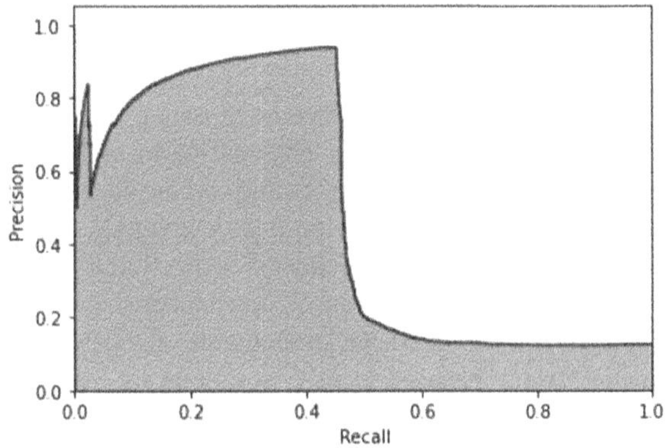

Fig. 3 Precision-recall curve with network flow features

Table 7 Descriptive statistics on features in the dataset

	Precision (%)	Number of PCs	Cutoff
Netflow	88.5	2	400
Biometrics	16.3	2	400
Netflow + biometrics	59.5	16	400

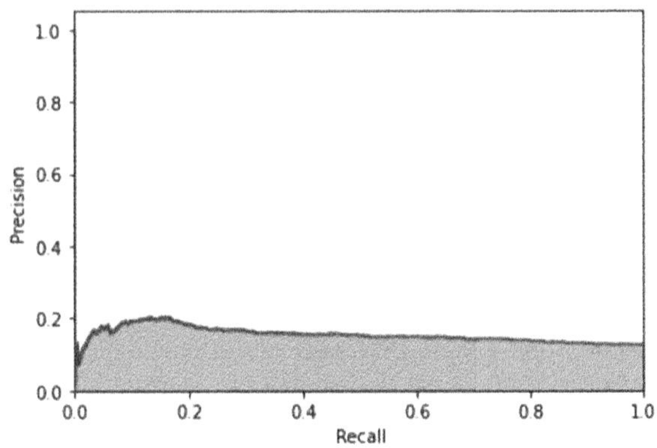

Fig. 4 Precision-recall curve with patients' biometrics features

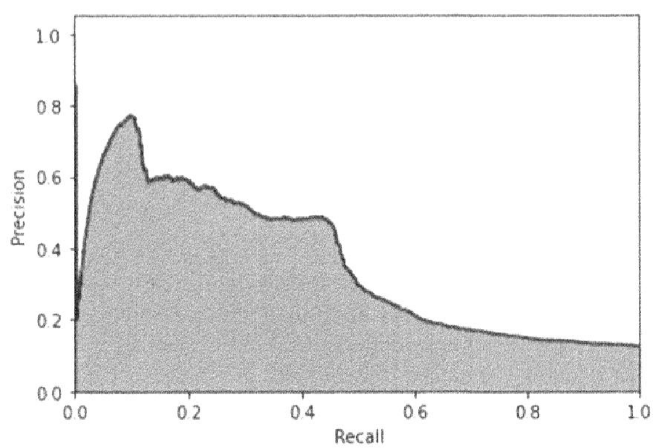

Fig. 5 Precision-recall curve with both patients' biometrics and network flow features

5 Conclusions and Future Work

Attacks on the mCPS devices have strong implications in the real world because these can cause potentially physical harm or life-threatening damage to patients, for example, overdosage of insulin when the medical insulin pumps are compromised, or a malfunctioned cardiac device (e.g. pacemaker) to endanger a patient's life. Existing security mechanisms in this domain include dedicated password-based authentication, access control mechanisms, encryption, and trust-based security management for wearable, implantable, environmental, and portal medical devices. Such prevention-based security solutions are highly insufficient mainly because (1) they can be easily circumvented by exploiting widely available vulnerabilities of embedded operating systems and wired/wireless communication protocols deployed in mCPS devices; (2) mCPS devices are usually considered as resource-constrained; implementing the computationally expensive security solutions inside the mCPS devices is challenging due to their limitations on power consumption and costly resources. In this chapter, we address this challenge from the perspective of the intrusion detection system with the technology of PCA-based anomaly detection in particular. The optimal attack detection accuracy the proposed approach can achieve is 88.5% by conducting an experimental evaluation with a publicly available dataset collected in an in-door testbed network.

In the future work, we will compare the current detection model with some other dimensionality reduction methods, such as sparse PCA, kernel PCA, Gaussian random projection and sparse random projection, with the same testbed benchmark. The current 88.5% detection rate is acceptable considering the system catches the malicious attacks without a prior knowledge on the attack type. To improve its performance further, we will investigate a hybrid approach integrating all types of dimension reduction based detection agents in our detection framework, including studies of theoretical foundation of detection [51], the architecture and implementation of the system [52], new performance metrics for evaluation criteria [53], the response to detected intrusions [54], traffic collection [55] and alert correlation to reduce false alarms [56].

Acknowledgements Research supported by New Hampshire- INBRE through an Institutional Development Award (IDeA), P20GM103506, from the National Institute of General Medical Sciences of the NIH.

References

1. Beasley RA (2012) Medical robots: current systems and research directions. J Rob 2012 (Article ID 401613). https://doi.org/10.1155/2012/401613
2. Rosen J, Hannaford B (2006) Doc at a distance. IEEE Spectr 43(10):34–39. https://doi.org/10.1109/MSPEC.2006.1705774

3. Rodrigues JJPC, Segundo DBDR, Junqueira HA, Sabino MH, Prince RMI, Al-Muhtadi J, De Albuquerque VHC (2018) Enabling technologies for the Internet of Health Things. IEEE Access 6:13129–13141
4. Papaioannou M, Karageorgou M, Mantas G, Sucasas V, Essop I, Rodriguez J, Lymberopoulos D (2020) A survey on security threats and countermeasures in Internet of Medical Things (IoMT). Trans Emerg Telecommun Technol. https://doi.org/10.1002/ett.4049
5. Islam SMR, Kwak D, Kabir MH, Hossain M, Kwak KS (2015) The Internet of Things for health care: a comprehensive survey. IEEE Access 3:678–708. https://doi.org/10.1109/ACCESS.2015.2437951
6. IBM Watson. https://cloud.ibm.com/developer/watson/dashboard. Retrieved 24 on Dec 2021
7. AWS. https://aws.amazon.com/. Retrieved on 24 Dec 2021
8. Microsoft Azure. https://azure.microsoft.com/. Retrieved on 24 Dec 2021
9. Gatouillat A, Badr Y, Massot B, Sejdić E (2018) Internet of Medical Things: a review of recent contributions dealing with cyber-physical systems in medicine. IEEE Internet Things J 5(5):3810–3822. https://doi.org/10.1109/JIOT.2018.2849014
10. Wang X, Wang L, Li Y, Gai K (2018) Privacy-aware efficient fine-grained data access control in Internet of Medical Things based fog computing. IEEE Access 6:47657–47665. https://doi.org/10.1109/ACCESS.2018.2856896
11. Yanambaka VP, Mohanty SP, Kougianos E, Puthal D (2019) PMsec: physical unclonable function-based robust and lightweight authentication in the Internet of Medical Things. IEEE Trans Consum Electron 65(3):388–397. https://doi.org/10.1109/TCE.2019.2926192
12. Cybersecurity Vulnerabilities Affecting Medtronic Implantable Cardiac Devices, Programmers, and Home Monitors: FDA Safety Communication. https://www.fda.gov/medical-devices/safety-communications/cybersecurity-vulnerabilities-affecting-medtronic-implantable-cardiac-devices-programmers-and-home. Retrieved on 24 Dec 2021
13. Makhdoom I, Abolhasan M, Lipman J, Liu RP, Ni W (2019) Anatomy of threats to the Internet of Things. IEEE Commun Surv Tutorials 21(2):1636–1675 (Second quarter 2019). https://doi.org/10.1109/COMST.2018.2874978
14. Zhang M, Raghunathan A, Jha NK (2014) Trustworthiness of medical devices and body area networks. Proc IEEE 102(8):1174–1188. https://doi.org/10.1109/JPROC.2014.2322103
15. Karageorgou M, Mantas G, Essop I, Rodriguez J, Lymberopoulos D (2020) Cybersecurity attacks on medical IoT devices for smart city healthcare services. In: IoT technologies in smart cities: from sensors to big data, security and trust; Institution of Engineering and Technology (IET), London, UK, pp 171–187
16. Lu W, Ghorbani AA (2008) Botnets detection based on IRC-community. IEEE GLOBECOM 2008—2008 IEEE global telecommunications conference, pp 1–5. https://doi.org/10.1109/GLOCOM.2008.ECP.398
17. Lu W, Mercaldo N, Tellier C (2020) Characterizing command and control channel of mongoose bots over TOR. In: Woungang I, Dhurandher S (eds) 3rd international conference on wireless, intelligent and distributed environment for communication. WIDECOM 2020. Lecture NOTES on data engineering and communications technologies, vol 51. Springer, Cham. https://doi.org/10.1007/978-3-030-44372-6_2
18. Lu W, Ghorbani AA (2008) Bots Behaviors vs. human behaviors on large-scale communication networks (extended abstract). In: Lippmann R, Kirda E, Trachtenberg A (eds) Recent advances in intrusion detection. RAID 2008. Lecture notes in computer science, vol 5230. Springer, Berlin, Heidelberg
19. Liu Z, Zhang L, Ni Q, Chen J, Wang R, Li Y, He Y (2018) An integrated architecture for IoT malware analysis and detection. In: Li B, Yang M, Yuan H, Yan Z (eds) IoT as a service. IoTaaS 2018. Lecture notes of the Institute for Computer Sciences, Social Informatics and Telecommunications Engineering, vol 271. Springer, Cham
20. Su J, Vasconcellos DV, Prasad S, Sgandurra D, Feng Y, Sakurai K (2018) Lightweight classification of IoT malware based on image recognition. In: 2018 IEEE 42nd annual computer software and applications conference (COMPSAC), pp 664–669. https://doi.org/10.1109/COMPSAC.2018.10315

21. Clincy V, Shahriar H (2019) IoT malware analysis. In: 2019 IEEE 43rd annual computer software and applications conference (COMPSAC), pp 920–921. https://doi.org/10.1109/COMPSAC.2019.00141
22. Tavallaee M, Lu W, Iqbal SA, Ghorbani AA (2008) A novel covariance matrix based approach for detecting network anomalies. In: 6th annual communication networks and services research conference (CNSR 2008), pp 75–81. https://doi.org/10.1109/CNSR.2008.80
23. Sampangi RV, Dey S, Urs SR, Sampalli S (2012) A security suite for wireless body area networks. arXiv:1202.2171, http://arxiv.org/abs/1202.2171
24. Lu W, Tavallaee M, Ghorbani AA (2008) Detecting network anomalies using different wavelet basis functions. In: 6th annual communication networks and services research conference (CNSR 2008), pp 149–156. https://doi.org/10.1109/CNSR.2008.75
25. Sangari AS, Manickam JML (2014) Public key cryptosystem based security in wireless body area network. In: 2014 international conference on circuits, power and computing technologies [ICCPCT-2014], pp 1609–1612
26. Lu W, Xue L (2) An enhanced CUSUM algorithm for anomaly detection. In: Traoré I, Awad A, Woungang I (eds) Information security practices, pp 83–96. Springer, Cham. https://doi.org/10.1007/978-3-319-48947-6_7
27. Li W, Zhu X (2014) Recommendation-based trust management in body area networks for mobile healthcare. In: 2014 IEEE 11th international conference on mobile ad hoc and sensor systems, pp 515–516. https://doi.org/10.1109/MASS.2014.85
28. Lu W, Tong H, Traore I (2008) E-means: an evolutionary clustering algorithm. In: Kang L, Cai Z, Yan X, Liu Y (eds) Advances in computation and intelligence. ISICA 2008. Lecture notes in computer science, vol 5370, pp 537–545. Springer, Berlin, Heidelberg. https://doi.org/10.1007/978-3-540-92137-0_59
29. Albalawi A, Almrshed A, Badhib A, Alshehri S (2019) A survey on authentication techniques for the Internet of Things. In: 2019 international conference on computer and information sciences (ICCIS), pp 1–5. https://doi.org/10.1109/ICCISci.2019.8716401
30. Trnka M, Cerny T, Stickney N (2018) Survey of authentication and authorization for the Internet of Things. Secur Commun Netw 2018 (Article ID 4351603):17. https://doi.org/10.1155/2018/4351603
31. Lu W (2005) A novel framework for network intrusion detection using learning techniques. In: PACRIM. 2005 IEEE Pacific Rim Conference On Communications, Computers And Signal Processing, pp 458–461. https://doi.org/10.1109/PACRIM.2005.1517325.
32. Lu W (2022) Cybersecurity data science: concepts, algorithms, and applications. In: Woungang I, Dhurandher SK (eds) 4th international conference on wireless, intelligent and distributed environment for communication. Lecture notes on data engineering and communications technologies, vol 94. Springer, Cham, pp 21–30
33. Lu W, Xue L (2022) A perceptron mixture model of intrusion detection for safeguarding electronic health record system. In: Barolli L, Chen HC, Enokido T (eds) Advances in networked-based information systems. NBiS 2021. Lecture notes in networks and systems, vol 313. Springer, Cham, pp 202–212
34. Nunley K, Lu W (2018) Detecting network intrusions using a confidence-based reward system. In: 2018 32nd international conference on advanced information networking and applications workshops (WAINA), pp 175–180. https://doi.org/10.1109/WAINA.2018.00083
35. "WUSTL EHMS 2020 Dataset". https://www.cse.wustl.edu/~jain/ehms/index.html retrieved on December 24, 2021
36. Ghorbani A, Lu W, Tavallaee M (2009) Network attacks, network intrusion detection and prevention: concepts and techniques. Springer, Berlin, pp 1–25. ISBN-10:0387887709
37. Garant D, Lu W (2013) Mining Botnet behaviors on the large-scale web application community. In: Proceedings of 27th IEEE international conference on advanced information networking and applications, Barcelona, Spain, March 25–28, 2013
38. Ghorbani A, Lu W, Tavallaee M (2009) Detection approaches, network intrusion detection and prevention: concepts and techniques. Springer, Berlin, pp 27–53. ISBN-10: 0387887709

39. Lu W, Ghorbani A (2008) Bots behaviors vs. human behaviors on large-scale communication networks. In: Lippmann R, Kirda E, Trachtenberg A (eds) Proceedings of 11th international symposium on recent advances in intrusion detection (RAID 2008). RAID 2008, LNCS 5230, MIT, Boston, USA, pp 415–416
40. Lu W, Miller M, Xue L (2017) Detecting command and control channel of botnets in cloud. Lecture notes in computer science (LNCS, vol 10618). Springer Nature, pp 55–62. ISBN 978-3-319-69154-1
41. Tavallaee M, Lu W, Ghorbani A (2009) Online classification of network flows. In: Proceedings of the 7th annual conference on communication networks and services research (CNSR 2009), Moncton, New Brunswick, Canada, May 11–13, 2009, pp 78–85
42. Lu W, Xue L (2014) A heuristic-based co-clustering algorithm for the internet traffic classification. In: 2014 28th international conference on advanced information networking and applications workshops, pp 49–54. https://doi.org/10.1109/WAINA.2014.16
43. Lu W (2007) An unsupervised anomaly detection framework for multiple-connection-based network intrusions. Ottawa Library and Archives Canada. ISBN 9780494147795
44. Lu W, Traore I (2005) A new unsupervised anomaly detection framework for detecting network attacks in real-time. In: Desmedt YG, Wang H, Mu Y, Li Y (eds) Cryptology and network security. Springer Nature, pp 96–109. ISBN 978-3-540-32298-6
45. Lu W, Traore I (2005) An unsupervised approach for detecting DDoS attacks based on traffic based metrics. In: Proceedings of IEEE pacific rim conference on communications, computers and signal processing (PACRIM 2005), Victoria, B.C., pp 462–465
46. Lu W, Traore I (2005) Determining the optimal number of clusters using a new evolutionary algorithm. In: Proceedings of IEEE international conference on tools with artificial intelligence (ICTAI 2005), Hongkong, pp 712–713
47. Lu W, Tong HJ (2009) Detecting network anomalies using CUSUM and EM clustering. In: Li Z (ed) Advances in computation and intelligence. Springer Nature, pp 297–308. ISBN 978-3-642-04843-2
48. Lu W, Traore I (2008) Unsupervised Anomaly detection using an evolutionary extension of K-means algorithm. Int J Inf Comput Sec 2(2):107–139 (Inderscience Publisher)
49. Lu W, Traore I (2005) A new evolutionary algorithm for determining the optimal number of clusters. In: Proceedings of IEEE international conference on computational intelligence for modeling, control and automation (CIMCA 2005), vol 1, pp 648–653
50. Shlens J (2014) A tutorial on principal component analysis. CoRR abs/1404.1100
51. Ghorbani A, Lu W, Tavallaee M (2009) Theoretical foundation of detection. In: Network intrusion detection and prevention: concepts and techniques. Springer, Berlin, pp 73–114. ISBN-10:0387887709
52. Ghorbani A, Lu W, Tavallaee M (2009) Architecture and implementation. In: Network intrusion detection and prevention: concepts and techniques. Springer, Berlin, pp 115–127. ISBN-10:0387887709
53. Ghorbani A, Lu W, Tavallaee M (2009) Evaluation criteria. In: Network intrusion detection and prevention: concepts and techniques. Springer Publisher, pp 161–183. ISBN-10: 0387887709
54. Ghorbani A, Lu W, Tavallaee M (2009) Intrusion response. In: Network intrusion detection and prevention: concepts and techniques. Springer, Berlin, pp 185–198. ISBN-10:0387887709
55. Ghorbani A, Lu W, Tavallaee M (2009) Data collection. In: Network intrusion detection and prevention: Concepts and Techniques. Springer, Berlin, pp 55–71. ISBN-10: 0387887709
56. Ghorbani A, Lu W, Tavallaee M (2009) Alert management and correlation. In: Network intrusion detection and prevention: concepts and techniques. Springer, Berlin, pp 129–160. ISBN-10:0387887709

Activity and Event Network Graph and Application to Cyber-Physical Security

Paulo Gustavo Quinan, Issa Traoré, and Isaac Woungang

Abstract The Activity and Event Network (AEN) is a new large graph model that enables describing and analyzing continuously in real-time key security relevant information about the operations of networked systems and data centers. The model allows identifying long-term and stealthy attack patterns, which may be difficult to capture using traditional approaches. The current chapter focuses on defining the model elements and the underlying graph construction algorithms, and presents a case study based on a cyberphysical security dataset.

Keywords Activity and event network · AEN graph · Graph model · Long-term attack · Attack patterns · Stealthy attack · Graph construction algorithm · Cyber-physical security · Probability model · Framework · Architecture

1 Introduction

Recently, it was discovered that a state-sponsored hacker group has been infiltrating the European Union's (EU) diplomatic communications network for years, downloading thousands of sensitive cables. The attack ran undetected for a three-year period and targeted more than 100 organizations and institutions, such as the United Nations and ministries of foreign affairs and finance. The attack is a type of emerging threat consisting of targeted and long-term campaigns delivered by skilled hackers who have clearly defined objectives and relentlessly work towards achieving their aims. These breaches can go undetected for a long period of time because of the hackers' ability to adapt to and escape defensive methods.

P. G. Quinan (✉) · I. Traoré
University of Victoria, Victoria, Canada
e-mail: quinan@uvic.ca

I. Traoré
e-mail: itraore@ece.uvic.ca

I. Woungang
Ryerson University, Toronto, Canada
e-mail: iwoungan@ryerson.ca

I. Traore et al. (eds.), *Artificial Intelligence for Cyber-Physical Systems Hardening*, Engineering Cyber-Physical Systems and Critical Infrastructures 2,
https://doi.org/10.1007/978-3-031-16237-4_10

Noticeably, there has been an evolution from volume-based attacks towards stealth like low and slow style attacks. Although volumetric attacks often occur within a set time frame, low and slow attacks rely on an ongoing stream of malicious requests and have no distinct beginning or end. This makes their detection by current Intrusion Detection Systems (IDSs) and Security Information and Event Management (SIEM) tools challenging.

The Activity and Event Network (AEN) graph model is a new security knowledge graph whose goal is to spearhead the development of a new generation of security data analytics techniques that can gain better situational awareness of the threat environment and allow detecting, responding and investigating sophisticated and stealthy attacks using data from both the traditional security ecosystem and beyond the organization perimeter. It leverages the large dynamic uncertain multigraph theory to coherently express and analyse security data across various heterogeneous data sources and meaningfully link seemingly innocuous and unrelated events to expose hidden and long-term attack patterns. The purpose of the current chapter is to define the AEN graph model elements and corresponding graph construction algorithms. Furthermore, a cyberphysical security dataset is used to illustrate some of the threat detection capability of the AEN model. One of the prime targets of long term attacks are cyber-physical systems, where quite often security is treated as an afterthought in system design and configuration.

The remainder of the chapter is structured as follows. Section 2 defines the theoretical foundation of the AEN graph model. Section 3 presents the data sources used to construct the graph. Section 4 defines the AEN graph model elements by presenting the node and edge types involved. Section 5 gives an outline of the AEN underlying probability model. Section 6 gives an overview of the AEN framework architecture and present a case study based on BoT-IoT cyberphysical security dataset. Section 7 makes some concluding remarks.

2 AEN Graph Theoretic Model

To start describing the AEN model, it is important to consider the most basic constituent elements of a network of interconnected devices like hosts, how they communicate with one another and that those elements have their own characteristics. The model, therefore, needs to incorporate those pieces of information. However, the data used to extract them might be incomplete, noisy or simply incorrect. That means the model must also be able to describe said uncertainty.

Also of importance is the network's dynamism: At any moment devices or hosts can be added or removed from the network; they can start or stop exchanging data; IP addresses can be recycled; devices can be infected and later be fixed, patched or updated etc. It follows that to be able to track the past existence of elements and their relationships through time, the model needs to maintain information on the time spans in which each element has existed. The dynamic graph model defined by Casteigts et al. [1] is used to formalize this control, with two changes: the first is that the model is expanded to combine probabilistic and temporal existence, while the second is the exclusion of latencies since they are negligible in the scope of this work. With that, important chronological relationships can be identified.

To summarize, the graph model has the following characteristics:

- Nodes have their own attributes. Thus nodes are labelled;
- Nodes can have multiple relationships at the same time. Thus nodes can have multiple edges between them;
- Relationships have a source and a destination. Thus edges are directed;
- Relationships have their own properties. Thus edges are labelled;
- Both relationships and nodes can be uncertain. Thus nodes and edges are weighed by probabilities of correctness, that is, that the information they represent is correct;
- Nodes and edges change through time and have an existence time span;
- Changes occur continuously;
- Processing times and latencies are negligible.

Those characteristics define the graph as a *Dynamic Uncertain Directed Multigraph*. Formally, it is defined as the 11-tuple

$$G = (N, E, s, t, \Sigma_N, \Sigma_E, \ell_N, \ell_E, \mathcal{T}, \pi_N, \pi_E) \tag{1}$$

where

- N is the set of all nodes of the graph as mentioned above;
- E is the set of edges representing relationships between nodes;
- $s : E \rightarrow N$, which assigns edges to their source nodes;
- $t : E \rightarrow N$, which assigns edges to their target nodes;
- Σ_N is a finite alphabet of available node labels;
- Σ_E is a finite alphabet of available edge labels;
- $\ell_N : N \rightarrow \Sigma_N$ is a map describing the labels of nodes;
- $\ell_E : E \rightarrow \Sigma_E$ is a map describing the labels of edges;
- $\mathcal{T} \subseteq \mathbb{R}^+$, that is, the time domain of the graph is in the positive real numbers;
- $\pi_N : N \times \mathcal{T} \rightarrow (0, 1]$, which assigns correctness probability values to nodes over time;
- $\pi_E : E \times \mathcal{T} \rightarrow (0, 1]$, which assigns conditional correctness probability values to edges over time given their endpoints.

3 AEN Data Sources

To construct the graph, heterogeneous data sources are used to extract data features capable of identifying nodes, relationships and their attributes. These involve both data sources available within the security perimeter as well as external data sources available through third-party services. Examples of internal data sources include network traffic logs, flow data, system logs, firewall logs, IDS alerts, anti-virus (AV) logs and email security logs, while examples of external data sources include Domain Name Server (DNS) queries and WHOIS. Of special note are the data sources which report security events or suspicious activity, like IDSes or SIEMs. Generically they are called *detectors* and their reports or logs are generically called *alerts*.

The features extracted from the data sources can be divided into two categories related to how they are collected:

1. **First-order** features, which are collected directly from the sources;
2. **Second-order** features which are derived from the available data either by actively using different tools and services or by mining them into aggregates.

First-order features can be further divided into two groups according to where they are sourced from:

- **Network**, which are the more traditional data used by IDS derived by analysis of the network traffic. Its features are: IP Address, Transport Protocol, Transport Port, and Content-Type.
- **Application**, which are the data collected from log analysis of known applications in the network and from service calls either to or from said applications in case those applications are programmed to provide such functionality. This would include data sources like hypervisor logs, syslogs and IDS alerts. Its features are: Application, User, Authentication State, Attack Type, Alert Confidence, and Device fingerprint.

Second-order features can also be further divided into two groups:

- **Third-Party**, which are collected by actively contacting external services or performing scans in order to collect more data about a certain entity. Its features are: *Domain name*, *IP Address*, *Name server*, *Autonomous System Number (ASN)*, *Location*, *Name*, *Email*, and *Operating System (OS)*.
- **Aggregates**, which are mined from the aforementioned features and from the model itself and may consist of temporal or spatial aggregates.

4 Graph Model Elements

In this section, we present the AEN graph model elements, specifically the node and edge types involved, and provide some illustrative examples. By analyzing the first and second-order features described in the preceding section, it is possible to identify some distinguishable characteristics in them that give clues into how the graph can be constructed. The first one is how each feature can be better used to model the network. Some of them, like IP addresses and domain names, can form relationships among themselves. These features are the nodes of the graph and their relationships are the edges. On the other hand, features like protocol and port are better employed as descriptors of said relationships and, therefore, are used to define attributes of either nodes or edges.

More generally, the nodes of the model are defined by any feature for which useful relationships could be formed, and the edges of the graph are defined by those relationships and their direction. The remaining features are used to define attributes of either nodes or edges.

4.1 AEN Nodes

Nodes can be classified into two distinct groups:

- **Active nodes**: Also called actors, active nodes are nodes that can be a source, a target, or a stepping stone (i.e., intermediary) in an attack. Examples of actors are hosts (either labelled by IP addresses or domain names), users (usually identified by authenticators like user name and email address) and devices. Active nodes can be further divided into:
  - *Internal actors*, which belong to or are under the control of the organization;
  - *External actors*, which are located outside the perimeter of the organization.
- **Passive nodes**: Nodes that carry some information or granular attributes for actors like DNS derived domain names and location nodes.

There are different types of nodes with each type having different features, as follows:

- **Account**: Represents an account in a system or application. It is derived from application and system logs and has the following properties:
  - Identifier
  - Application
- **Alert Group**: Aggregation of related detector alerts, called raw alerts internally, that might be generated in a short time frame thus associated to a single event or attack. It has the following properties:
  - Protocol
  - Source IP
  - Source port
  - Destination IP
  - Destination port
  - Service
  - Classification
  - Start time
  - Stop time

- Severity
- Confidence
- Alert count

- **Domain**: Represents the domain name of an IP address. It is derived from a reverse DNS lookup of an IP address. When an IP address is identified, a DNS Reverse DNS lookup cycle is formed until the complete Domain/IP relationship is found. This is useful to identify attacks using Fast-Flux DNS and Algorithmically Generated Domains (AGDs). It has the following property:
  - Name
- **Host**: Represents a network host that communicates with other hosts in the network. It is an active node type and it is the most important node type of the model, from which almost all other elements originate. It is derived from different data sources like network data (packets or flow data), logs and alerts. It may be labelled with host-specific aggregates and historical information like if it was ever part of an attack, etc. It has the following property:
  - Identifier: Piece of information that can be used to consistently and uniquely identify a host though time, like IP address, MAC address, fingerprint or just a generic identifier value. Note that the IP address is not the ideal identifier for this case but given its ubiquity, it is used when other identifiers are not available.
- **IP Address**: Represents an individual IP address. It can be a first-order feature, derived from network data, logs or alerts, or a second-order feature when it is derived from DNS lookups or Network Address Translation (NAT) table conversions. It has the following property:
  - IP address
- **IP Range**: Represents a range of IP addresses. It is derived from WHOIS queries and has the following property:
  - Classless Inter-Domain Routing (CIDR) block
- **Location**: Represents a geographical location. It is derived from WHOIS queries or from IP geolocation services. It is useful to correlate hosts from similar locations and attribute attacks. It has the following property:
  - Tag: a combination of city, state/province, region and country according to what is available on the result of the query
- **Organization**: Represents an organization which controls a range or IP addresses or owns domain names. It is derived from WHOIS queries and has the following property:
  - Name

- **Person**: Represents a person who is listed as being an administrator or owner of a range of IP addresses, organization or domain names. It is derived from WHOIS queries and has the following properties:
  - Name
  - Email

4.2 AEN Edges

Each node type has different types of edges originating from and terminating in them. The different types of edges and their features (when applicable) are as follows:

- **Authentication Attempt** $Account \rightarrow Host$: Represents an authentication attempt of an account into a host. It has the following properties:
  - Timestamp
  - Source IP
  - Successful: Whether the authentication was successful or not
- **Triggered By** $AlertGroup \rightarrow Host$: Describes the source of an alert group
- **Used** $Host \rightarrow IPAddress$: Expresses that a host used an IP address to send data
- **IP Located At** $IPAddress \rightarrow Location$: Links an IP address to its location as defined by the WHOIS or geolocation query performed when the IP address was first added to the graph or when the relationship was considered stale
- **Part Of** $IPAddress \rightarrow IPRange$: Links an IP address with its IP range as defined by the WHOIS query performed when the IP address was first added to the graph or when the relationship was considered stale
- **Resolved To** $IPAddress \rightarrow Domain$: Links an IP with its domain name as returned by the reverse DNS query performed when the IP address was first added to the graph or when the relationship was considered stale
- **Controls** $Organization \rightarrow IPRange$: Expresses that an organization controls an IP Range as described by the WHOIS queries performed when the IP Range was first added to the graph or when the relationship was considered stale
- **Organization Located At** $Organization \rightarrow Location$: Links an IP address to its location as defined by the WHOIS query performed when the first IP belonging to the organization was first added to the graph or when the relationship was considered stale
- **Owns** $Person \rightarrow Domain$: Represents an ownership relationship between a person and a domain as described by the WHOIS query performed when an IP that resolved to this Domain was first added to the graph or when the relationship was considered stale
- **Session** $Host \rightarrow Host$: Represents a communication session between two hosts on the same protocol, ports and within a certain activity time window. Its direction

is based on who initiated the connection. It is primarily an aggregation of network data, but data from other sources like logs and alerts that can be used to identify such communication is also used in order to fill up possible gaps in the network data available. It has the following properties:

- Start time
- Stop time
- Protocol
- Source port
- Destination port
- Source size: Sum of the length of all packets from source to destination
- Destination size: Sum of the length of all packets from destination to source
- TCP state: Only applicable for TCP packets, it describes the state of the TCP connection as of the last packet belonging to this session, that is, if the connection has been established (handshake completed), finished or has only ever started but not been established
- Packet count: The number of packets exchanged between the hosts as part of the session
- Fragmented packet count: The number of fragmented packets exchanged between the hosts as part of the session
- Alert count: The number of alerts generated as part of the session

5 AEN Probability Model

5.1 Probability Model Definition

The basic probability assignment involved in the AEN model consists of the probability of correctness, π, for graph elements (nodes and edges) and feature confidence. At inception, each graph element is assigned an correctness probability which stems from the originating data.

Most graph elements are derived from data that in some way can be categorized as deterministic; like when packets related to a TCP handshake between two hosts are received, the model can be certain that those two hosts are communicating and in what direction, or when an application reports that an authentication attempt was made for a certain account there is no doubt regarding the application or the account being used. In these cases, π is trivial and set to 1, or more technically to $1 - \epsilon$, where ϵ represents the inherent probability the data injected into the system has been faked, tampered or corrupted.

On the other hand, a few data sources are inherently imperfect, which introduces a layer of uncertainty to the nodes, relationships and attributes they generate and thus must be taken into consideration. Examples of these data sources include IDSes or other detectors, user/device fingerprinting schemes, geolocation services etc. More

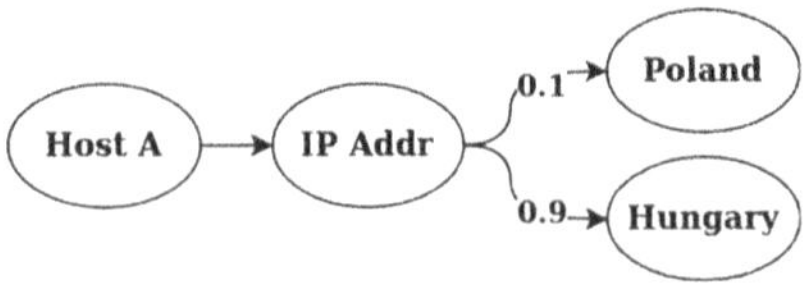

(a) Graph representing two possible locations for an IP address node with different probabilities

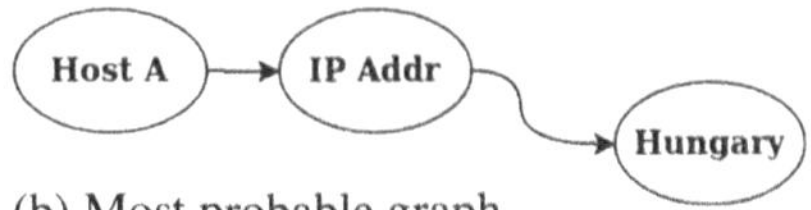

(b) Most probable graph

Fig. 1 Example of probability modelling in the AEN graph and the resulting most probable subgraph derivation

than that, in some cases the data have a higher probability of being faked or spoofed, like a stray IP packet not part of a established communication between two hosts which is more likely to have a spoofed source address. In these cases, the correctness probability for the graph elements will correspond to the expected accuracy of the data.

As an example, consider the hypothetical scenario where the geolocation service assigns for a given IP address 90% probability of it being from Hungary and 10% probability of it being from Poland at a given point in time $t \in \mathcal{T}$. Based on that, one *IP Address* node and two *Location* nodes will be added to the graph. The two countries exist beyond any doubt so their π are set to 1. Likewise, the IP address, at least as a member of the set of all possible IP addresses, also exists, and is therefore also assigned $\pi = 1$. From there, one edge between the *IP Address* and each of the two *Location* nodes constrained to t are also added with their probability of correctness being set to the respective probability returned by the geolocation service. In such a manner, both possible, but conflicting, relationships can be modelled and, at a later point, used as part of different inference processes.

Figure 1 depicts a sample graph based on the above scenario with (a) showing the full graph representing the 2 conflicting relationships and (b) showing the resulting "most probable graph" derivation. Future data might result in updated values and thus different derivation results.

In the above scenario, all probabilities are equivalent to probabilities of existence, however, that is not always the case. Consider now an alert added to the graph. Short of an attacker being able to inject a false alert into the system, the model can be sure that the alert exists and was generated by the detector. Therefore, its probability of existence is equal to $1 - \epsilon$. That, however, doesn't properly represent the uncertainty regarding the alert. Instead, what is important in this case is describing the probability of the alert being correct, that is, not a false alarm.

Generally, a detector can be considered as a binary classifier that classifies events as either malicious (or at least suspicious) or benign and generates alerts when an event is classified as malicious. As a classifier, the detector has an accuracy which underpins the probability of correctness of the alerts it generates. Moreover, some detectors will also provide a confidence score for the alert which can be used to further refine the alert's probability.

Therefore, to calculate π for an alert, we first need to define its expected accuracy. There are different levels of granularity that can be used when performing the accuracy calculation. The coarsest grain (and simplest) calculation would be to calculate the precision (Positive Predicted Value—PPV) of the detector as a whole by evaluating it against different datasets and use that as the surrogate. The precision is used in this case because it describes the ratio of true positives among all predicted positive elements, in other words, it is congruent to the probability of a predicted positive observation (an alert in our case) being a true positive.

This is simple but the values obtained may be too generic and thus not as useful. A finer grained approach would be to group alerts based on a common factor and calculate the accuracy for each group separately. The grouping can be done in several different ways like by severity, by family of attacks, by individual attack type inside a family or even by individual rules or anomaly metrics. These would provide more specific probability values but on the other hand would require much more data in order to be calculated meaningfully. If there are no examples or not enough examples of a given alert in the datasets, then the respective values cannot be calculated using this approach or at least not calculated meaningfully. In these cases, the general detector accuracy would have to be used.

In cases where a confidence score is available, we compute the alert probability by combining the detector accuracy and the confidence score into a probability value through score calibration. The score calibration process is described in more detail in a separate report [10].

Finally, recall that alerts are not added directly to the graph, instead, similar alerts of the same type, origin and target that occur in a short time frame are grouped into alert groups. In most cases, all alerts of the same type will have the same π, hence, the probability of any alert can be used as the probability of the group. The exception to that is cases where different alerts have different confidence scores. In those cases, to obtain π of an alert group A from it's constituent alerts A_i we pick the probability of the alert with the highest confidence. Formally, $\pi(A)$ is defined as:

$$\pi(A) = \max_{i=1}^{n} \pi(A_i) \tag{2}$$

The probability of correctness serves as a starting point to inferences, derivation or other types of analyses in the graph as seen in the following sections.

5.2 Probability Model Usage and Application

It is expected that the basic probability assignment (i.e. alert probability) will be fed to the model by leveraging the threat detection systems (IDSes, AVs, anti-phishing systems) already available in the organization. However, this is not required. Regardless of whether there are some pre-existing IDSes in the organization, the AEN will provide independently its own threat detection schemes which encompass attack fingerprinting, graph clustering and unsupervised statistical threat detection and that will be fed back to the model. As a result, the AEN threat detection scheme will also provide some of the classification schemes mentioned in the above probability model. As new threats are discovered by the AEN threat detection schemes, alerts will be generated and incorporated in the graph model along with the alerts generated by other pre-existing schemes if applicable. Although the AEN threat detection schemes could leverage the alerts generated by pre-existing schemes, it is our goal, currently, to keep them separate to ensure the independence of the AEN schemes. In future work, we will explore how pre-existing alerts information can be leveraged by the AEN threat detection schemes for detection purpose.

Currently, we use the AEN probability model mainly for threat assessment and visualization, and providing underlying context and explanatory information. This is based on the concept of *threat horizon* and *reverse threat horizon*.

The horizon of node u can be understood to be the set of all possible nodes which u could have affected or exchanged data with. In the case of an attack, all the nodes for which a direct journey exists are nodes that the attacker could have compromised directly, even if only temporarily. On the other hand, nodes for which only indirect journey exists can only have been compromised if somewhere along the journey the attacker was able to permanently compromise the system either through some kind of outbound connection to its servers or through the installation of an automated malware.

That demonstrates the importance and capacity of the horizon in regards to the forensic analysis of the network. It provides a complete view of the network from the point of view of the originating node and defines what could possibly be within its reach should it be a threat to the network. In other words, the horizon can be viewed as the subset of nodes which can be threatened by another node.

Under this optics, the horizon of a node u is here defined as its ***Threat Horizon***, denoted as $\overrightarrow{TH_u}$, and defined as the subset of N in which all elements are reachable from u. Formally:

$$\overrightarrow{TH_u} \subseteq N, \quad \overrightarrow{TH_u} = \{w \in N : u \rightsquigarrow w\} \tag{3}$$

Inversely, the ***Reverse Threat Horizon*** of a node v, denoted as $\overleftarrow{TH_v}$, identifies, from the point of view of a target node, which network nodes could have attacked and compromised it. It is defined as the subset of N in which all elements can reach v. Formally:

$$\overleftarrow{TH_v} \subseteq N, \quad \overleftarrow{TH_v} = \{w \in N : w \rightsquigarrow v\} \tag{4}$$

The Threat Horizon can be used as a starting point of any node specific analysis. The first step is to select the best source node for the Threat Horizon, which is characterized by its power to link distinct attacks together. Therefore, the selection, when available, of owners, common domains or users might result in a combined Threat Horizon of multiple attacks.

The same applies for identifying the target node of the Reverse Threat Horizon. Generically, those nodes are called the **focal points** of analysis.

By using the basic probability model defined above for the AEN model we can derive probability measures for the threat horizon and reverse threat horizon, and use these values to guide threat assessment, visualization, and forensic analysis. This is an ongoing work that will be presented in more details in future papers.

6 Graph Construction and Framework Implementation

6.1 Framework Architecture

The system architecture of the AEN framework is depicted in Fig. 2.

The central component of the system is the AEN Engine, which is responsible for maintaining the AEN graph. It provides key functionalities like processing and aggregating incoming data through data receivers, attack fingerprint matching, proactive third-party data collection to supplement other incoming data, anomaly detection and the probabilistic model underlying the AEN graph.

The engine stores the graph in a custom-made, in-memory, graph database with capabilities to add, update, remove and search for graph elements. Persistence is obtained via frequent storage of snapshots which can be reloaded from the disk in case of a failure or to review a previous graph state. These snapshots can also be shared with other tools and systems that provide other auxiliary operations or functionalities like visualization of the graph or scalability via read replicas.

Both the engine and the graph database are implemented in Java and are executed in a single process which allows for direct memory access of the graph elements, thus avoiding any extra serialization overhead.

The engine provides two operation modes. The first one is the online mode, in which real-time data is continuously collected from external machines and devices by a client application which then pre-processes and sends the live data to the engine.

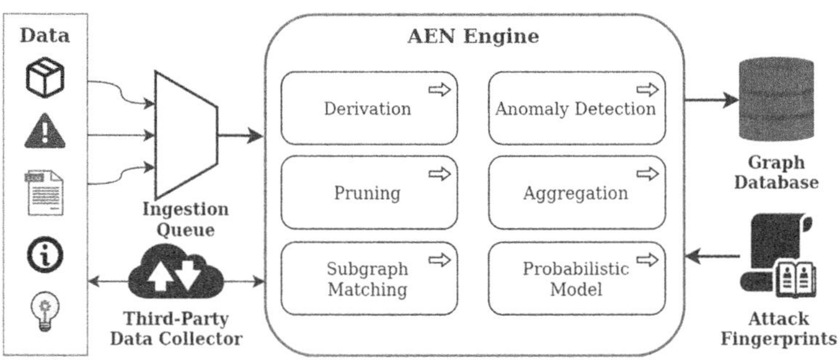

Fig. 2 AEN system architecture

Table 1 Bot-IoT files used in experiment

Type	Name
Data exfiltration	IoT_Dataset_data_theft__00002_20180618111101.pcap
Data exfiltration	IoT_Dataset_data_theft__00013_20180618112736.pcap
OS scan	IoT_Dataset_OSScan__00001_20180521140502.pcap
OS scan	IoT_Dataset_OSScan__00003_20180521150020.pcap
Service scan	IoT_Dataset_ServiceScan__00007_20180515133133.pcap
Service scan	IoT_Dataset_ServiceScan__00007_20180521224912.pcap
UDP DDos	IoT_Dataset_UDP_DDoS__00019_20180604180729.pcap

The second one is the offline mode, in which previously collected data is added to the graph directly. In both cases, AEN Engine uses data ingestion queue which sorts the data chronologically and controls the flow of data.

The current implementation supports network traffic data, either raw or flow data, some syslogs, IDS alert logs from Snort, Zeek and custom alerts derived from Kitsune's anomaly detection output and pre-collected IP addresses information derived from DNS and WHOIS queries, which the engine can also actively query for if the data is not available.

6.2 Case Study Based on a Cyperphysical Security Dataset

To better demonstrate the functionalities and capabilities of the AEN Graph, we performed an experiment using the BoT-IoT dataset [2–7], provided by the Cyber Range Lab of the University of New South Wales (UNSW) Canberra, which consists of legitimate and simulated Internet of Things (IoT) network traffic, along with different botnet attacks.

The graph was generated using a subset of the dataset with 1.4 GB of pcap data comprising part of the available UDP-based DDoS, OS scans, service scans and data exfiltration attacks. Table 1 lists the selected files. This data resulted in a graph with 236 nodes, including 56 hosts, and 337, 349 edges, including 337, 073 sessions.

Figure 3 shows a zoomed out visualization of most of the resulting graph. Blue nodes are hosts while the edges between them are sessions. To help visualization, multiple edges between the same nodes are combined into a single, thicker, edge. Note how the graphical representation of the data clearly shows a cluster of activity around a few hosts with high inter-connectivity and outside of that a host with high centrality that initiated connections to several different hosts. It's those kind of patterns that can be identified by matching algorithms to expose otherwise hard to identify relationships between hosts and anomalous or known malicious behaviours.

By zooming in and adding labels to the graph elements, it is possible to see in more detail how the different node types are interconnected through the different edges. Figure 4 shows that visualization.

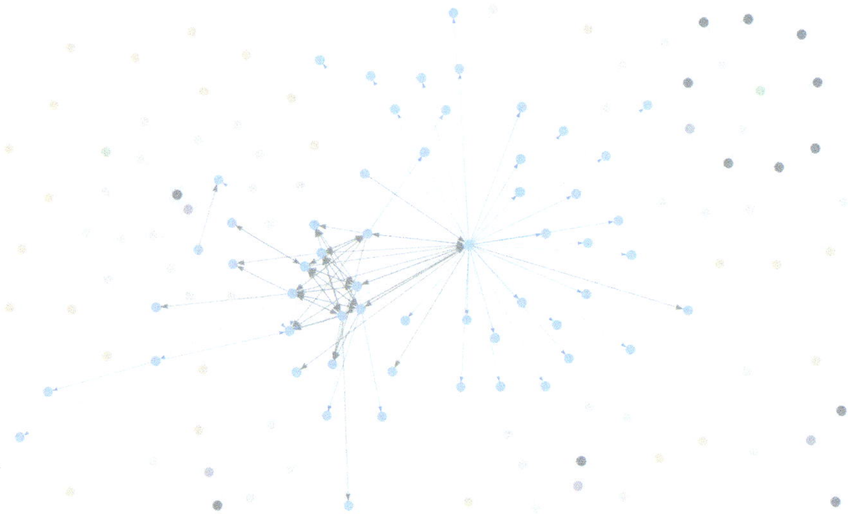

Fig. 3 Zoomed out AEN graph sample derived from the BoT-IoT dataset

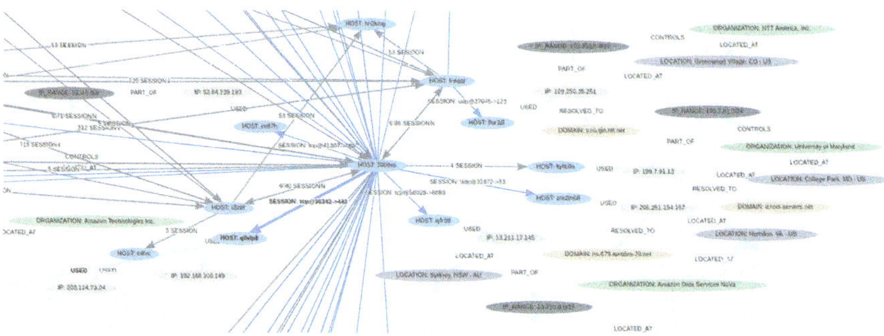

Fig. 4 Zoomed in AEN graph sample derived from the BoT-IoT dataset

To label the graph, we derived host labels from the dataset's flow-level labels such that any host is labelled as malicious if it originated any flow labelled as being part of an attack. Out the 56 hosts in the graph, 8 are labelled as malicious, a prevalence of 14.3%.

As a base of comparison, we first ran the data used to build the graph through Snort IDS using the rules available on its website for registered users. To classify hosts, we followed a similar rule based on the Snort alerts, with hosts that were the source of any alert being classified as malicious by Snort.

The results show a high number of errors, with it only being able to correctly classify 4 out of the 8 malicious hosts and 34 out of the 48 benign hosts. A high number of type II errors is expected given that Snort is a signature based IDS but on the other hand, the high number of false positives (type I errors) was not expected for the same reason. Table 2 presents the resulting confusion matrix.

Table 2 Confusion matrix of the Snort IDS

		Predicted	
		Malicious	Benign
Actual	Malicious	4 (50%)	4 (50%)
	Benign	14 (29%)	34 (71%)

Table 3 Confusion matrix of the AEN's anomaly detection algorithms

		Predicted	
		Malicious	Benign
Actual	Malicious	8 (100%)	0 (0%)
	Benign	4 (8%)	44 (92%)

More specifically, the performance of the Snort IDS for the dataset is thus: Sensitivity of 50%, specificity of 71%, precision of 22% and negative prediction rate of 89%. That gives a F1-score of 0.3 and a Matthews Correlation Coefficient (MCC) of 0.15.

Afterwards, we executed the anomaly detection algorithms of the AEN [9] against the generated graph and followed the simple classification rule in which each host identified as anomalous was classified as malicious.

The anomaly detection algorithms were able to identify all 8 hosts as anomalous but also identified further 4 benign hosts as anomalous. Table 3 presents the resulting confusion matrix.

Compared to the results obtained by Snort, our algorithms exhibited a much higher performance with a sensitivity of 100%, specificity of 92%, precision of 67% and negative prediction rate of 100%. Furthermore, the F1-score is 0.8 and the MCC is 0.78.

Finally, we also performed matches against our attack fingerprint database as described in [8]. As in the other cases, host classification adhered to the simple rule where a host was classified as malicious if it was reported as matching by the detector. In this case, that meant the host was identified as being the source of an attack pattern that matched any of the stored fingerprints.

Like Snort, the fingerprint were able to correctly classify the same 4 out of the 8 malicious hosts. However, it did not misclassify any of the 48 benign hosts, which is expected given that (a) it is by nature signature based and (b) only a few fingerprints are available. Table 4 presents the resulting confusion matrix.

Compared to the results obtained by Snort, the fingerprint matching also showed a much higher performance with a sensitivity of 50%, specificity of 100%, precision of 100% and negative prediction rate of 92%. Furthermore, the F1-score is 0.67 and the MCC is 0.68.

Table 4 Confusion matrix of the fingerprint matching

		Predicted	
		Malicious	Benign
Actual	Malicious	4 (50%)	4 (50%)
	Benign	0 (0%)	48 (100%)

When compared with the anomaly detection, it showed a better false positive rate compared to a worst false negative error which is natural given their different characteristics and shows they can be complimentary.

7 Conclusion

The AEN graph model is a new paradigm that allows the capture and analysis of the activities and events involved in the operation of networked systems and data centers. In the current chapter, we have defined the graph model elements and provided an overview of the underlying probability model. While the focus of the current chapter is on graph construction only, future papers will present in more detail our approaches for threat detection using the AEN graph model.

References

1. Casteigts A, Flocchini P, Quattrociocchi W, Santoro N (2012) Time-varying graphs and dynamic networks. Int J Parallel Emergent Distrib Syst 27(5):387–408
2. Koroniotis N (2020) Designing an effective network forensic framework for the investigation of botnets in the Internet of Things. Ph.D. thesis, UNSW Canberra
3. Koroniotis N, Moustafa N (2020) Enhancing network forensics with particle swarm and deep learning: the particle deep framework. In: International conference on artificial intelligence and applications
4. Koroniotis N, Moustafa N, Schiliro F, Gauravaram P, Janicke H (2020) A holistic review of cybersecurity and reliability perspectives in smart airports. IEEE Access 8:209802–209834
5. Koroniotis N, Moustafa N, Sitnikova E (2020) A new network forensic framework based on deep learning for internet of things networks: a particle deep framework. Futur Gener Comput Syst 110:91–106
6. Koroniotis N, Moustafa N, Sitnikova E, Slay J (2018) Towards developing network forensic mechanism for botnet activities in the iot based on machine learning techniques. In: Hu J, Khalil I, Tari Z, Wen S (eds) Mobile networks and management. Springer International Publishing, Cham, pp 30–44
7. Koroniotis N, Moustafa N, Sitnikova E, Turnbull B (2018) Towards the development of realistic botnet dataset in the internet of things for network forensic analytics: Bot-iot dataset
8. Nie C, Quinan PG, Traoré I, Woungang I (2022) Intrusion detection using a graphical fingerprint model. In: 2022 22nd IEEE international symposium on cluster, cloud and internet computing (CCGrid), pp 806–813

9. Quinan PG, Traore I, Gondhi UR, Woungang I (2022) Unsupervised anomaly detection using a new knowledge graph model for network activity and events. In: Renault E, Boumerdassi S, Mühlethaler P (eds) Machine learning for networking. Springer International Publishing, Cham, pp 117–130
10. Yousef WA, Traore I, Briguglio W (2022) Classifier calibration: with application to threat scores in cybersecurity. IEEE Trans Dependable Secure Comput, pp 1–1